U0174368

食 物 小 传

白兰地

Brandy

A Global History

〔美国〕贝基·苏·爱泼斯坦 著

陈媛熙 译

北京联合出版公司
Beijing United Publishing Co.,Ltd.

目录

序　言　奢华干邑，迷人白兰地　　　　　　　　　　　　　　1

第一章　炼金术：从古典文明到干邑　　　　　　　　　　　001

第二章　白兰地的生产：蒸馏与陈酿　　　　　　　　　　　009

第三章　干邑享誉全球　　　　　　　　　　　　　　　　　021

第四章　雅文邑及其光辉历史　　　　　　　　　　　　　　031

第五章　欧洲和高加索地区的著名白兰地　　　　　　　　　045

第六章　西班牙和拉美的杰出白兰地　　　　　　　　　　　061

第七章　澳大利亚和南非　　　　　　　　　　　　　　　　075

第八章　美国出产的白兰地　　　　　　　　　　　　　　　085

第九章　干邑知识大百科　　　　　　　　　　　　　　　　093

第十章　干邑鸡尾酒与21世纪的趋势　　　　　　　　　　105

第十一章　小批量生产的白兰地和干邑　　　　　　　　　　113

鸡尾酒配方　　　　　　　　　　　　　　　　　　　　　　123

致　谢　　　　　　　　　　　　　　　　　　　　　　　　137

序言

奢华干邑，迷人白兰地

> 不，先生，红葡萄酒是给毛头小子喝的；男子
> 汉要喝波特酒；但谁若是渴望成为英雄（微微一笑），
> 那就非喝白兰地不可。首先，白兰地的味道能让味
> 蕾得到无上享受；其次，人在酒中所能得到的一切，
> 白兰地顷刻之间便可给予。

——詹姆斯·鲍斯威尔，《塞缪尔·约翰逊传》（1791 年）

如今，白兰地又重回聚光灯下。它是鸡尾酒会的时尚宠儿，是盛装在上千美元雕花水晶瓶中的高端啜饮酒，也是名流聚会中的常见饮品。然而，就在几十年前，白兰地还远未达到如此显赫的地位。

人人都知道什么是白兰地——不过真的是这样吗？白兰地经葡萄酒蒸馏而成，是一种口感美妙、气味芳香的烈酒。由于经年累月储藏在木桶中陈酿，酒液通常会呈现出琥珀或桃木的色泽。今天，在美国和英国最负盛名的白兰地当属干邑和雅文邑，它们分别出产于法国南部的两个地区。

西班牙白兰地和其他种类的白兰地在世界各地也广受欢迎。

几十年前，价格不菲的白兰地只有富裕人家才消费得起，而且也只是在餐后惜酒如金地小酌几口；上了年纪的人可能会在餐柜深处存放一瓶普通白兰地，指望它发挥一些似有若无的药效；也有一些白兰地是作为廉价酒出售的。然而对于大多数人来说，白兰地同他们的生活完全没有交集。

世界上最著名的白兰地——干邑，一直是一个亮点。产自法国干邑地区的白兰地声望始终如日中天。当被问到对干邑的看法时，大多数人哪怕从未接触过这种酒，也会对它颇具好感。干邑是最著名也是最昂贵的白兰地，这一观念被时下的明星们一再强化。富有传奇色彩的电影导演马丁·斯科塞斯曾亲自担纲主演轩尼诗系列广告；说唱明星卢达·克里斯拥有自己的干邑品牌"魔术"；史努比·狗狗曾为朗帝干邑代言；还有许多成功的说唱歌手把自己最喜欢的干邑写进他们的热门曲目。在如今的高端酒吧，技艺高超的酒保（现在叫调酒师）会在调制干邑鸡尾酒时各显身手、一较高下，他们会选取精妙的风味配料调配出口味卓绝的混合饮品，将这种优雅芬芳的烈酒凸显得越发醇美。

严格来说，白兰地可以用各种水果蒸馏制成，但考虑到本书的主题，我们就暂且将白兰地定义为用葡萄酒制成的烈酒。葡萄酒被蒸馏成烈酒后，通常会储存在木桶中陈酿，白兰地那种动人的黄褐色就是在此过程中形成的；这才是本书所指的"白兰地"。

虽然年份较短的白兰地可以用来制作鸡尾酒或助人放

松解闷儿，但啜饮陈年干邑或白兰地才是真正值得回味的难忘经历。当一口酒精度达40%的上好烈酒流经体内时，我们就会体验到一种美妙的舒适和幸福感。（也许这就是药用疗效？或许吧。）

20世纪末，白兰地的运势陷入低谷，是干邑使它重新焕发生机。在此之前，出身干邑地区的高档白兰地一直被视为年长成功人士（主要是男性）的专属饮品。直到20世纪中叶，人们仍然认为女性在公共场合享用干邑有失体面，除非是为了搭配特定菜肴，比如橙香火焰可丽饼——盘中以白兰地为主料的酱汁在餐桌上被点燃，这道甜点便绽放出炫目的火焰。在世界各地，许多财力有限的人也习惯饮用白兰地，不过这类白兰地价格却要便宜许多，大部分是品质稍逊的本土品牌，用葡萄酒（以及其他原料）制成，但仍被冠以"白兰地"甚至"干邑"之名。白兰地在人类文明史上经历了漫长的发展历程，期间这样的高低起落一直在上演，而这段演进之旅的起点需要追溯到700多年之前。

从中世纪开始，白兰地就被写进药方，用于治疗多种疾病和症状。到了现代，白兰地成了许多种潘趣酒①的基础配料，这种酒曾在18、19世纪的美国和英国大受追捧。19世纪中叶，美国掀起了有史以来第一波鸡尾酒热潮，白兰地又成了调制鸡尾酒的烈性基酒。

白兰地对蒸馏工艺流程要求非常严格，因此这种烈酒

---

① 潘趣酒是一种特色混合性饮料，属于软性饮品或微酒精饮品，主要成分是果汁。——本书注释，除特别说明，均为译者注

的生产成本并不低，加之还要在酒桶中经过多年陈酿，一瓶酒的价格自然会更加昂贵。干邑、雅文邑与赫雷斯白兰地是旧大陆最著名、最昂贵的三种贵族白兰地。在南非和澳大利亚历史早期，白兰地产业在当地蓬勃兴旺，因为它们乃是英荷两国的殖民前哨，而这两个国家的人一向有消费白兰地的习惯。如今，由葡萄酒制成的高品质白兰地在亚美尼亚、格鲁吉亚以及美国等国家已有100余年的生产历史。

由于干邑是最广为人知的奢华白兰地，许多国家便仿效起干邑地区的制酒方法，其中大部分竟然将自产白兰地也称作"干邑"。于是，正名问题便逐渐浮现在法国干邑生产商面前，自19世纪末全球白兰地产量激增以来，他们一直在积极解决这一问题。

同样是在19世纪末期，葡萄根瘤蚜大爆发令欧洲葡萄园遭受重创，法国优质干邑的产量开始下降。到20世纪初，鉴于干邑价格高昂、供应有限，许多流行鸡尾酒便使用威士忌和其他烈酒来代替白兰地。因此，对于大多数消费者而言，白兰地（尤其是干邑）与其日常生活的距离益发遥远。这种情形一直持续到20世纪中叶，当时，图书和电影里出现的无数场景都强化了一种印象：饮用白兰地的人往往有种自命不凡的优越感。1971年，风度翩翩的特工詹姆斯·邦德在白兰地的世界里掀起了一场小小的风暴：在电影《007之金刚钻》中，他随手从会客厅拿起干邑，将它带入火线，而此处的"火线"为字面意思——他将一瓶拿破仑干邑用

收获期进入尾声时，干邑地区的葡萄园开始褪去丰沛的绿意，呈现出干邑白兰地本身的棕褐色和琥珀色。

作武器去火攻对手。不过总而言之，似乎白兰地已经逐渐沦落到被人们放在日益闲置的会客厅里。

然而，到了 20 世纪末 21 世纪初，白兰地（确切地说是干邑）突然重见天日，被带进了主流社会。促成这一变化的主要是名为"都市音乐"的流派和一些出人意料的代言人——说唱和嘻哈艺人。20 世纪下半叶，低端白兰地在美国的某些城市地区非常流行。都市音乐风格大火之时，就连并不住在这些地区的年轻人也开始模仿自己在广播中听到、在视频中看到的生活方式。随着嘻哈和说唱明星人数与日俱增，拿破仑、轩尼诗以及其他顶级品牌的奢华瓶装干邑在其歌曲和音乐视频中的出场率也越来越高，粉丝们热血沸腾，追随偶像一起涌入奢华干邑的世界。干邑的消费量骤然飙升。

与此同时，酒保们也不断翻新花样，逐渐成为专业的"调酒师"，鸡尾酒在城市高档场所的人气也越来越高。虽然我们不能就此断言音乐或调酒师就是终极催化剂，但结果已清晰明了：白兰地（尤其是干邑）王者归来，重领风骚。

如今，在纽约和其他国际化大都市，人们通常是在晚餐之前而不是之后喝白兰地。傍晚，吧台后方的调酒师开始施展自己的魔法，将干邑倒入鸡尾酒调酒壶中，再加入精选的风味配料以及其他烈酒和利口酒，调制出绝妙的冰爽鸡尾酒。越来越多追求刺激的酒客在初次品尝白兰地（尤其是干邑）时就采用这种方式：在高档酒吧中，将它与其他酒兑在一起，做成酒力劲爆的混合饮品。价格昂贵的干邑鸡尾酒已在美国大城市率先蹿红，而如今在国际范儿调酒师的引领下，全球其他城市也不甘落后，开始追逐这一诱人潮流。

英美两国向来是陈酿白兰地消费的领跑者。然而，消费量猛增十余年之后仍未出现衰减迹象的中国在这方面正逐渐赶超西方。基础款白兰地在全球依然有庞大的市场，马来西亚、菲律宾、印度等亚洲国家都是白兰地的主要消费国。这种烈性酒缘何能在这么多的层面上广受欢迎？这些层面分别又是哪些呢？

除了原产地之外，对白兰地价格和品质影响最大的就

最近，哈迪干邑酒庄推出了一系列干邑，以雕刻华美瑰丽的玻璃酒瓶盛装。这些干邑恰巧呼应了干邑本身的炼金术起源，因为它们是以炼金术中的四大元素来命名的，即土、气、火、水。哈迪干邑还加入了第五个"元素"——光。

是陈酿。但所有的白兰地都始于同一道工艺流程——蒸馏。蒸馏技术从古文明世界的中心被带到欧洲后再进入新世界，这是一段历时数世纪的漫长旅程，也是我们马上将要探索的故事。

蒸馏设备（用于水的蒸馏）

［第一章］

炼金术：从古典文明到干邑

火焰与黄金，狄俄尼索斯①与克利奥帕特拉②，早期基督教与秘密教派都在蒸馏的历史中发挥了作用。古典文明时期，最早使用蒸馏法的人目的各异。有人是为了寻找长生不老药，或者说"生命之水"。还有人将水与火这两种截然对立的元素奇迹般地融为一体，生产出 aqua ardens③：这是一种神奇的"燃烧之水"，是可燃的液体。后来，到了中世纪早期，炼金术士试图炼制所有金属中最贵重的一种——黄金，在此过程中制造出了不同种类的馏出物。远在公元前 1 世纪的克利奥帕特拉统治时期，早期的埃及人就开始探索蒸馏法，当时一些深受尊敬的哲学家兼化学家所采取的两种蒸馏法大同小异。信仰狄俄尼索斯的希腊人也会使用蒸馏葡萄酒。

　　罗马帝国成为基督教国家之后，蒸馏法似乎已销声匿迹，但实际上它只是隐入了地下。秘密教派诺斯底派利用这种工艺为自己的神圣仪式制备液体长达数百年；在整整一千年的历程中，虔诚的清洁派教徒都要经历一次真正的"火之洗礼"，其中就要用到以蒸馏法制成的"燃烧之水"。这项技术或许还传到了亚洲，不过也有可能亚洲在同一时期发明了这项技术。据说早期的阿拉伯炼金术士曾听过这样的传言：公元 4 世纪时，中国的道士们就会用蒸馏法来炼制"不老仙丹"。然而在 3 世纪的西方，即便是普通水手也

---

① 希腊神话中的酒神。
② 克利奥帕特拉七世（公元前 69—前 30 年），被称为埃及艳后，是古埃及的托勒密王朝最后一任女法老。
③ 拉丁语，直译为"燃烧的水"，即烈性酒。

懂得蒸馏的概念：基本上就是先蒸发海水，再小心地将其冷凝，从而得到漫漫航程中迫切需要的脱盐淡水。对他们来说，这种能让他们活下来的水是名副其实的"生命之水"。

此时，距离将蒸馏法用作武器仅有一步之遥了。公元672年左右，在马尔马拉海域的基兹科斯战役中，交战一方的水手祭出了"燃烧之水"。在战场上，拜占庭人向来袭的萨拉森战船投掷燃烧的液体，成功守住了君士坦丁堡这座伟大城市。虽然这种"希腊火"中除了蒸馏酒之外，说不定还添加了石油（也有可能只有石油，没有蒸馏酒），但不管怎样，蒸馏法及其产物"燃烧之水"的威名从此远播四方。

关于蒸馏的知识还经由中东传播到了波斯帝国，波斯人借助蒸馏法，在新兴的草药学上取得了更多成果。公元6世纪，波斯国王库思老一世在贡德沙普尔城兴建了一所医学院，环绕学院的园圃遍植草药、花卉和其他种类的植物。在这里，蒸馏酒精发挥着关键作用，因为所有药用浸剂的制备都会用到它。

摩尔人在占领西班牙后实行禁酒令，但原有的葡萄园并未因此被毁，因为在制造香水和化妆品的蒸馏过程中也要用到葡萄。事实上，人们认为 alcohol（酒精）一词就是来源于阿拉伯语 kohl，意思是黑色眼影粉（在阿拉伯语中，al 的意思是"这个"或"一个"，最终组合成单词 al-kuhl[①]）。阿伦比蒸馏器中的 alembic（阿伦比）一词同样

---

① 在阿拉伯语中，指通过蒸馏而得到的精华。

源于阿拉伯语，而阿拉伯语中相应的单词又来源于希腊语ambix，意为"杯子"。早在1265年，alembic这个单词就在法语书面语中出现过。

在欧洲，蒸馏是炼金术士把普通物质炼制成黄金的一道工序。事实上，一种包含蒸馏的化学过程的确能给某些物质镀上金色的涂层，这似乎足以激励一代又一代人为了造出真金而想方设法完善这种工艺。人们不仅将黄金视为最重要的金属，还相信它是对付各种病症的灵丹妙药。（格但斯克金箔酒是现代德国出产的一种利口酒，其商标名Goldwasser的字面意思就是"黄金水"，令人不禁联想到这种观念。）

中世纪时期，随着化学、炼金术和医药学彼此交织融合，蒸馏法在所有这些领域都占据了一席之地。到中世纪晚期，经过炼金术士和医师们的推广，用于制备药物的蒸馏法已传遍整个北欧和不列颠群岛。然而这时，这个世界又跃跃欲试，准备将其另作他用——我们或许可以称之为休闲？

不过，蒸馏葡萄酒的结果在早年间却是喜忧参半的。例如，为了卖出更多的葡萄酒，德国商人在蒸馏过程中会放入一些内含有毒化学物质的添加剂。许多人尝试用葡萄酒之外的其他原料进行蒸馏，但结果五花八门，大多数很糟糕。最终，这种实验也促成了威士忌的问世，这当然算是个好结果。然而在此之前，我们就已经造出了白兰地。

通过蒸馏葡萄酒就能靠谱地生产出可饮用的白兰地。当时，这是一种清澈透明、未经陈酿的烈酒，其最早的商业用途之一就是强化已酿成的葡萄酒。人们将烈酒兑到葡萄酒中，通过增加其酒精含量来稳定酒液，防止变质。有时，为了提高葡萄酒的甜度，人们也会在发酵过程中加入烈酒，这样在酵母将葡萄中的糖分全部转化为酒精之前，烈酒就会把酵母杀死。如今，产自法国地中海沿岸露喜龙地区的上等甜葡萄酒仍然使用当地蒸馏的烈酒进行强化。

炼金术符号包含四种基本元素：土、气、水、火。炼金术士常常施展出高超本领，包括操控元素和驱策象征意义极强的动物，如本图中的蛇形龙。

蒸馏技术继续一路北上，13世纪便已传入法国加斯科涅地区，有可能是朝圣者从圣地亚哥—德孔波斯特拉①带回来的。有证据表明，1299年教皇克雷芒五世的私人医生阿诺·德维尔纳夫曾用蒸馏葡萄酒制备的药物为他治病。医生把这叫作"生命之水"，拉丁语为 aqua vitae，用法语说则是 eau de vie。时至今日，法国人仍将其用作蒸馏烈酒的通称。

不久之后，在位于加斯科涅的雅文邑地区，人们开始将这种葡萄酒的蒸馏液储存（陈酿）在木桶中，就此揭开了白兰地的现代篇章。在700多年前的1310年，正是雅文邑地区用自己的名字为第一批陈酿白兰地命名。以加斯科涅为新起点，白兰地蒸馏法继续传播：向南直达西班牙安达卢西亚地区，也就是赫雷斯镇的所在地（赫雷斯不仅是雪利酒的故乡，也是佳酿赫雷斯白兰地的原产地）；向北沿法国大西洋海岸一直传到波尔多、干邑和卢瓦尔河谷。

从16世纪初期开始，荷兰贸易商就沿欧洲的大西洋海岸往来航行。他们会专门将法国的葡萄酒运回本国，这是因为荷兰气候过于寒冷，不适宜种植酿酒葡萄。与大多数白兰地简史中所记载的相反，干邑地区出产的蒸馏烈酒并没有立刻迅速地传播开来。事实上，荷兰商人采购葡萄酒的地点是一些小城镇，它们坐落在流入法国西部大西

---

① 西班牙加利西亚自治区的首府。相传耶稣十二门徒之一的雅各安葬于此，是天主教朝圣胜地之一。自中世纪以来，前来此地的朝圣者络绎不绝，乃至形成了一条有名的朝圣之路，即圣雅各之路。

洋海岸的河流沿岸。最初荷兰商人蒸馏一部分低度葡萄酒，也许是为了使其在归途中保持稳定，也许只是因为将葡萄酒蒸馏浓缩后再运输效率更高。这种酒最初的名字叫brandewijn（荷兰语意指煮制或"烧制"的葡萄酒），后来才简称为brandy（白兰地）。

最晚从 1536 年起，荷兰人就开始喝白兰地了，相关记载见于一条禁止酒馆老板出售外带白兰地的法规。英国曾从国外进口白兰地供居民制作利口酒，"强化"低度葡萄酒，或用作草药和香料的基本原料。但是，17 世纪末爆发的

荷兰的船只非常适于航海，时常穿梭往来于欧洲大西洋沿岸的水域；图中的这艘船在大海中劈波斩浪，一路上抓紧时间采购各色货品，例如干邑地区的蒸馏葡萄酒，将它们带回荷兰卖给翘首以待的顾客。

"光荣革命"和九年战争使英国、法国及荷兰的干邑生意遭受重创。冒牌白兰地的生意自然趁机大行其道——人们用各种水果和香料制成蒸馏酒，企图仿造出法国白兰地的味道。

另一个因17、18世纪政治局势而获利的"产业"就是走私。不过，即便在没有战争的时候，走私白兰地也不愁销路。它们从英国沿海的多处水湾入境，以此来逃避关税。爱尔兰的上流社会人士也爱喝白兰地，甚至会将手下的一部分商人差往干邑地区经营相关业务——这也是为什么如今某些举世闻名的干邑酒庄会拥有一个爱尔兰名字，比如轩尼诗和豪达（Otard，最初叫作 O'Tard）。

在欧洲和英美两国，人们曾认为白兰地可供药用，因此大多数家庭都会将它作为常备药，用来治疗从昏厥到消化不良等各种症状。旅行者会随身携带白兰地，靠它来补充体力、杀菌消毒。如今许多人都曾听说或使用过"药用白兰地"这个词，却并不了解它的确切含义。在过去的几个世纪里，无论是被人视为生活必需品还是休闲饮品，这种名叫白兰地的蒸馏葡萄酒一直是整个西方世界必不可少的重要商品。

[第二章]

白兰地的生产：蒸馏与陈酿

大家都知道，白兰地是一种金棕色的烈酒。当真如此吗？大多数消费者并不熟悉白兰地的生产工艺，甚至连鉴酒行家也概莫能外。白兰地最初其实是无色透明的液体。大部分白兰地都是在木桶中陈酿，而木桶会赋予这澄澈液体迷人的琥珀色，并使其随着时间的推移颜色逐渐加深。但无论透明还是有色，这种烈酒始终是白兰地。事实上，西方调酒界新近流行起来的一种烈酒，就是名为皮斯科的无色白兰地。没错，原产于秘鲁的皮斯科的确是白兰地的一种。它的制作方法同法国干邑和西班牙赫雷斯白兰地是一样的：用葡萄酒蒸馏而成。但它通常不会在橡木桶中陈酿，因此颜色才会清亮透明。

本书重点讨论的葡萄酒蒸馏白兰地是三大品种白兰地之一，另外两种是用其他水果和果渣蒸馏出来的白兰地。以水果为原料制成的著名白兰地包括苹果白兰地、李子白兰地和樱桃白兰地，不过在全球各地，一些国家和区域也对许多其他水果蒸馏出来的白兰地情有独钟。果渣白兰地是用酿酒葡萄的皮、梗和籽蒸馏而成，这些原料都是葡萄经压榨酿酒后余下的残渣。两种最著名的果渣白兰地是意大利的格拉帕和名字起源于法国的玛克。我们不在本书中介绍水果白兰地和果渣白兰地的原因是，大多数人认为，像格拉帕这样的烈酒不能算是白兰地。再说，放眼世界，美妙绝伦的葡萄白兰地实在是太多了。

这三种白兰地的共同点是皆以蒸馏法制成。经过蒸馏，它们就从原料（葡萄、水果、果渣）转化成了烈酒。回顾

刚刚蒸馏出来的白兰地是一种无色透明液体，放入木桶中熟成后，就会慢慢呈现出浓郁的金黄色和琥珀色。

蒸馏的历史，早在白兰地开始成为众所周知的可贸易商品之前，中世纪的炼金术士就在运用这种工艺流程了。spirit 这个词无论是在酒类的语境下，还是在哲学和宗教学科中，都是指某种强劲而富有活力的物质或本质[1]，或许这并非偶然。我们已经在前文获知蒸馏法是如何从中东一路传入欧洲的，现在，了解一下这道工艺在不同地区白兰地生产中的应用情况应该会颇具妙趣。

---

[1] 在英语中，spirit 这个单词既可以指烈酒，也有精神、心灵、灵魂、精髓等含义。

简单地说，蒸馏就是在密闭容器中加热葡萄酒，然后收集由此产生的芳香乙醇蒸汽，最终将其冷却为液体的过程。将葡萄酒加热到沸点，在这个温度下，酒精的蒸发量大于水（酒精的沸点是78摄氏度，即172华氏度，而水的沸点是100摄氏度，即212华氏度）。酒精被浓缩于最终形成的冷凝液体：此时，葡萄酒就变成了烈酒。酒精在蒸发和冷凝的过程中会携带葡萄酒的一些香气和味道，因此，每种白兰地都有其独一无二的特色。

当然，蒸馏白兰地并不仅仅是把葡萄酒煮沸那么简单；如果真是这样，那么任谁都能做到。在实际操作中，整个过程可以说是非常复杂而美妙：在一间宽敞的屋子里，一只锃亮的铜质容器如同一个巨大的水壶在闪闪发光，房间里弥漫着铜壶下方木柴燃烧产生的烟雾。这只大铜壶通过倾斜的、圆筒状甚至螺旋形的管子同其他容器连接起来，整套装置看起来仿佛一部散发出俏皮光芒的大型实验室设备——尤其是当你想到它所造之物是何等美妙时，这装置就更显得赏心悦目。

影响白兰地酿造的主要因素有三个：葡萄酒的品种、蒸馏设备的类型和蒸馏师的技术。每个白兰地产区也都各有一套决定其最终产品品质的规则和传统，其中的要素包括可以使用哪种蒸馏器，必须在一年当中的哪几个月进行蒸馏，以及关于陈酿工艺和酒瓶形状的要求。

大部分用来蒸馏白兰地的葡萄酒，都是由具备两种特质的白葡萄酿造而成的。首先，这些葡萄内含有效的酸性成分；其次，它们能在蒸馏后赋予白兰地理想的香气和味道。从传统上来讲，用于制造白兰地的葡萄是酿不出上佳葡萄酒的，不过这条规律也并非放之四海而皆准。

葡萄采收后，经发酵酿成葡萄酒，随后必须立即进行蒸馏，这一点非常重要（除非葡萄酒可以在控温环境中保存），因为常用的葡萄酒防腐剂二氧化硫（SO_2）实际上会毁掉蒸馏的烈酒。所以，历史上干邑和雅文邑地区都曾规定，葡萄酒的蒸馏必须始于某个特定的日子（就在葡萄刚刚充分发酵成葡萄酒之后），止于冬季结束之前，趁着天气依然寒冷，尚能以天然方式来储存葡萄酒。现如今，葡萄酒在用于蒸馏之前，可以一直在冷藏罐中保存数周或数月，但这项传统却在各产区制定的规则中延续下来。

最常见的做法是，在蒸馏前，首先将葡萄酒中的酒泥（发酵过程中沉淀下来的酵母残渣和葡萄酒中的其他残留物）沥出，因为酒泥很容易掉落到蒸馏器底部燃烧起来，把整批白兰地都毁掉。不过，也有些酿酒商会在蒸馏时保留酒泥，他们说这样生产出来的酒更加美味，为此值得多费些功夫，在蒸馏器中安装螺旋桨式搅拌装置，让酒泥在葡萄酒中不断循环流动。

蒸馏某一产年全部葡萄酒的过程可能需要持续好几个月，因为在任何一家酿酒厂，蒸馏器和蒸馏大师的数量通常都是有限的。在白兰地酿造行业，蒸馏工艺基本上分为两种：一次蒸馏法和二次蒸馏法。多次蒸馏法目前正在伏特加行业推广和普及，因此你或许会认为蒸馏次数多多益善，但这一点并不适用于白兰地。随便找一位来自雅文邑的人问问，你就明白了，在当地，一次蒸馏法才是他们尊崇的传统。然而，如果移步到干邑地区，你可能又会觉得二次蒸馏才是不二法门。

这两种蒸馏工艺都会产生许多杂质，从各种散发恶臭或味道骇人的东西（比如杂醇油，尝起来更像是发动机里用的，而不是入口喝的）到极端危险的化学物质，不一而足。"酒头"[1]是比乙醇更易挥发的化合物，这意味着它们的蒸发和冷凝温度都比乙醇低；"酒尾"[2]则是挥发性弱于乙醇的化合物，其蒸发和冷凝的温度比乙醇高。在二次蒸馏法中，通常要在第二次蒸馏开始和结束时将酒头和酒尾丢弃。干邑和其他一些地区都采用二次蒸馏法，也就是说葡萄酒要经过两次蒸馏过程；有时在第二次蒸馏中会加入部分酒头或酒尾，有时则不会，具体如何选择取决于公司的做酒传统和首席蒸馏师的风格。

一次蒸馏法的拥趸（比如雅文邑和赫雷斯地区的酿酒师）会告诉你，这种工艺的要求远比二次蒸馏法更为严格。在一次蒸馏法中，蒸馏塔会利用温度梯度将酒头与酒尾从达到标准的烈酒中分离出来。首席蒸馏师必须精心把控塔式蒸馏器的温度和运转，以便去除多余的化合物，并在一次蒸馏过程中保留所有必要的香气、味道和酒精度。

许多蒸馏师会使用燃气来加热蒸馏器，但也有一些酿

---

① 最先从甑锅流出的酒，酒精度数可达70多度，醛类物质多，暴烈味大。
② 最末从甑锅流出的酒，酒精度数低于10度，酸类物质多，邪杂味大。

依据传统，雅文邑的蒸馏过程在塔式蒸馏器中进行，有时这种设备非常小巧，甚至可以放在四轮马车上从一个农场运到另一个农场。

酒厂选择用木柴生火加热。如果游客在秋季蒸馏期造访雅文邑的塔西克等酒庄以及干邑和赫雷斯地区，就会幸运地闻到烟熏的香气，这会带给他们更加美妙的体验。

蒸馏过程结束后，产出的烈酒就可以陈酿了。在桶陈过程中，酒液中会沁入各种迷人的香气，如香草味、太妃糖味、果干味，以及雪松和香料的气味。在大多数地区，陈酿期即将结束时通常要向酒液中加入一定量经过严格过滤的清水，将白兰地稀释到最终可供饮用的强度，即酒精度40%左右。大多数白兰地的"桶强"（出桶时的酒精浓度）约为60%至70%，不过这一数值会随陈酿年数变化，因为在陈酿过程中酒精和水都会蒸发（且速度不一）。

赫雷斯白兰地是西班牙的名贵白兰地，它与干邑采用的是同一套传统酿制工艺。图中是西班牙赫雷斯地区百艾斯酒庄里一些年代久远、使用频繁的铜质壶式蒸馏器。

每个地区在制作用于陈酿白兰地的酒桶时，都有自己偏爱的木材。干邑附近的利穆赞橡树被认定为黄金标准，不过，在雅文邑和高加索山脉等地，也生长着木材质地纹理与之类似的橡树林。有些白兰地生产商甚至会亲赴山林挑选用于制作酒桶的树木，还有些厂商可能会到自家制桶工场里去挑选已经切割好的木料。制作陈酿桶的木材必须先放置数年，使其自然风干。

干邑和其他高端白兰地的陈酿桶大多是手工制作的。首先，稍稍加热已风干的桶板使其柔韧，然后围着一小堆火，将它们□□□□□□□花朵的样子。在干邑地区，这个过程叫作□□□□□□□□想是做成玫瑰的形状。然后，将桶箍套□□□□□□□慢慢敲击，使其下降到合适的位置，将□□□□□□□木桶内部被"烘烤"到位后，将桶底装□□□□□□完毕的木桶滚动运出烟雾氤氲的高温□□□□□流程已经有几百年的历史了。

木桶□□□□□□□□以设在地上，也可以深入地下。白□□□□□□□□能会被挪动位置，比如从温暖之处□□□□□□从湿度较大的地方搬运到湿度较小□□□□□□产作坊会把酒桶存放在自家农场的□□□□□□多小作坊是这样做的，而大型干邑□□□□□□其他类型的仓库里。

但是□□□□□□□□里都有一种相同的东西——□□□□□□□直到不久前，陈酿干邑的木桶□□□□□□条具有柔韧性。不过，干邑地区有一种小虫嗜吃柳木，而陈酿室里的蜘蛛则以小虫为食，因此能让干邑酒桶在数十年内保持完好，所以干邑生产商都会仔细留心不去惊扰这些蜘蛛。即便如今大多

在干邑地区，储存陈酿桶的酒窖并不一定设在地下。它还可以是某个小农场中的一座石砌谷仓，那里的空气甚少被扰动，而蜘蛛则在干邑熟成的过程中守护着酒桶。

数酒桶都已改用金属桶箍固定，但仍有不少人迷信地认为，一个健全的酒窖里必须有大量蜘蛛——这或许同干邑的奢华形象反差巨大，但却至关重要。

传统干邑酒桶是用柳条捆扎固定的。

干邑享誉全球

在参观干邑市那座小小的艺术博物馆时，你不可能忽视一个细节：无论是在肖像画还是雕像中，500年前的法兰西国王弗朗索瓦一世似乎总是面带微笑。他完全有理由开心。在他统治期间，河畔小镇干邑已经是该地区远近驰名的航运中心，白兰地生意也开始萌芽。荷兰商人正在当地拓展业务，寻找新的贸易产品。

关于干邑白兰地的"发明"，流传最广的说法是：为便于运输，荷兰商人对当地葡萄酒进行强化，待返回家乡后再兑水将其还原。但这只是真实历史的简化版本。在当时那个年代，荷兰商人是最精明的贸易商。他们眼光精准独到，知道要买什么、卖什么，造什么、如何造，才能获得最多的利润。自16世纪起，他们就孜孜不倦地沿法国大西洋海岸开展贸易。他们带回荷兰的重要商品之一，就是购自法国沿海贸易港口拉罗谢尔的盐。拉罗谢尔和雷岛之间只隔着一条窄窄的海峡，后者至今仍以品质上佳的盐而闻名。此外，拉罗谢尔还恰好位于流经干邑镇的夏朗德河河口附近，因此这里自然也能买到葡萄酒。

在通往大西洋的航线上，荷兰人在卢瓦尔河畔的大型贸易城市南特建立了一处商业殖民地，卢瓦尔河谷的酿酒厂就将自家出产的葡萄酒运至此处。在更南边的夏朗德河地区，荷兰人也建了一处殖民地。他们还从大西洋驶入吉伦特河口湾，一路航行至波尔多开展贸易——这是一段轻松通畅的旅程。

由于其中一些法国葡萄酒口味很淡且容易变质，为了便于运输，贸易商就想办法使其保持稳定。他们还希望提高葡萄酒的陈酿潜力——不过并不是像我们今天所想的那样，让葡萄酒在瓶中陈酿多年直至味道成熟。在当时那个年代，葡萄酒只要能在一年的陈酿期内保持不变质即可，也就是坚持到次年葡萄收获被酿成葡萄酒为止。

但事实证明，卢瓦尔河谷和波尔多出产的葡萄酒被蒸馏成烈酒之后，反倒没有原来值钱了。因此，从卢瓦尔河谷和波尔多进口的葡萄酒，荷兰人会直接卖出去，而干邑出产的葡萄酒却仍是蒸馏后再销售，因为它们的利润更高。

17世纪也是一个政治激变和宗教动荡频仍的时代，这对干邑白兰地的贸易也造成了影响。英法之间的战争和冲突还在继续，而在法国、英国、荷兰和其他欧洲北方国家，天主教徒和新教徒在地方、国家和国际领域的斗争中彼此厮杀，打得难解难分、乱作一团。

17世纪发生的一场重要局部战争对干邑跃升为地区经济霸主产生了重大影响。1651年的一次战役中，在城墙环绕的干邑城里，居民们成功阻击了与波旁王室亲王路易二世为敌的军队，而当时亲王是路易十四军中的一位将军。为表感谢，国王后来免除了干邑地区葡萄酒和蒸馏酒的税收及关税。凭借这一财政优势，干邑地区的经济发展很快开始超越周边，成为这片区域内所有本地出口产品（包括干邑白兰地）的贸易中心，干邑白兰地也正逐渐成为世界上最负盛名的白兰地。

欧洲和英国的政治斗争一直持续到 18 世纪。为了不再仰赖法国人提供他们所钟爱（同时也获利颇丰）的烈酒，一些英国人便来到夏朗德地区自建干邑酒庄。时有发生的英法冲突还让爱尔兰人乘虚而入进军干邑产业。

如此一来，荷兰商人在法国白兰地贸易中就丧失了近乎垄断的地位。荷兰人仍然嗜饮烈酒，但此时白兰地已经不是他们唯一的选择了：荷兰等北方国家由于气候过于寒冷，无法种植酿酒葡萄，但蒸馏技术的进步已经使其居民能够在当地用谷物酿造烈酒。

包括白兰地在内，所有这些烈酒最初都被称为"生命之水"。但它们很快就拥有了自己的专属名称，有的是取自当地语言，有的是以其产地命名。荷兰人了有荷式金酒（现代金酒的前身）；苏格兰人开始酿造苏格兰威士忌；俄国和其他欧洲北方国家生产起了伏特加；朗姆酒从新大陆渡海而来。但在这些国家的上层社会，尤其是在荷兰、德国和英国，人们最爱的仍然是干邑。18 世纪上半叶，法式、英式和爱尔兰式名号开始出现在干邑酒庄的门面上。时至今日，其中一些酒庄仍然是顶级干邑生产商，比如马爹利、轩尼诗和人头马。

与此同时，干邑开始同其他北欧烈酒（即"生命之水"）拉开了距离。没错，它们全都是激情似火的烈酒，是人生一大美妙享受，是炼金术、医学和宗教实践交相融合的产物。但相较于当时的其他"生命之水"而言，干邑的口感更为柔滑细腻一些，味道也更醇美。

还有几个因素也一并为干邑白兰地的质量和声誉锦上添花：其一，干邑地区的地理位置靠近利穆赞森林，制作上等干邑陈酿桶的木材即产自此处；其二，干邑先是在巴黎流行，随后风靡法国在全球的殖民地，其相应的时间段也很重要。到 18 世纪初，法国北部的居民已经对干邑白兰地有了更多了解，但与英国人不同的是，他们并不是非高端啜饮酒不喝。在极端天气导致葡萄园歉收、地产葡萄酒供不应求的年份，他们就开始喝普通干邑。其实，干邑在法国的声威素来不及在其他国家，部分原因是法国人很早就饮用干邑，还有部分原因在于干邑主要被视为出口产品。

用于出口的干邑自然要被装进以本地木材制作的木桶中。即便是在中世纪，干邑产区周围的森林也是大名鼎鼎的，后来森林中的树木还被路易十四的海军使用过。利穆赞地区出产的这种橡木在今天也堪称品质一流。在利穆赞橡木（其树种为英国橡树，也叫法国橡树，即夏橡）制成的桶中，干邑能够充分陈酿，因为这种树的木质相对松散，正好能让适量的酒液渗入桶壁，使烈酒在木桶中获得恰到好处的颜色、辛香和味道。

在木桶中陈酿过的干邑会呈现出琥珀般的色泽，这进一步将干邑与同期以其他水果和谷物酿就的"生命之水"区分开来。其他那些烈酒往往是当地自产自销，因未经桶陈，颜色清澈透明。很快，木桶陈酿就成为干邑地区（正如在雅文邑一样）烈酒的一个关键特征，酿酒商们熟练掌握木桶的制作和养护技艺，只为提升自家干邑的品质。

当木桶陈酿成为干邑的一项重要特征后，"陈酿"这一概念本身也承载了全新的意义。术语"拿破仑干邑"就是这样应运而生的。根据目前干邑地区的法律，拿破仑干邑必须符合特定的陈酿要求，但这只是干邑地区近期颁布的规定，并没有在全世界普遍施行。

著名干邑酒厂库瓦西耶[①]的口号是"拿破仑之干邑"。库瓦西耶最初是一家位于巴黎市郊的葡萄酒和烈酒公司，据说拿破仑·波拿巴曾于1810年御驾亲临此地。或许正因如此，在旷日持久的拿破仑战争期间，这位皇帝突然开始每天早晨向军队发放定量干邑来鼓舞士气。库瓦西耶自豪地将自己与陛下的关系广而告之。几年后，公司决定专注于干邑的生产，于是在1828年将总部迁至干邑地区。直至今日，库瓦西耶的总部仍驻于当地，就在夏朗德河畔干邑上游的雅纳克镇。

有人断言，正是因为拿破仑力捧干邑，（在干邑和世界其他地区）才会有更多的白兰地生产商开始以这位皇帝之名来命名自己的干邑。有人则认为，更重要的因素是19世纪以来陈酿干邑地位上升，于是更多的厂商用"拿破仑干邑"一词来指代据说是自拿破仑时代就开始陈酿的名贵干邑。无论是哪种情况，"拿破仑"这个名字既显著提高了干

---

① 库瓦西耶是Courvoisier的音译，为酒庄创始人的姓氏，但由于该酒庄同拿破仑的渊源，加之品牌徽标上印有拿破仑的剪影，因此中文一般将这种酒直接译为"拿破仑"，酒庄也被称为"拿破仑干邑酒庄"，但同干邑等级中的"拿破仑"并不是同一概念。

邑的声望，也大幅增加了它的销量，以至于"拿破仑"之名经常被别国的其他白兰地盗用，这是干邑地区生产商至今仍要应对的问题。

随着干邑重要性凸显的时代渐渐来临，19世纪40年代，英国首相罗伯特·皮尔将干邑的关税削减了近三分之一，助它一举突破保护主义关税的难关。到了1860年，关税进一步降低，六年之后，输入英国的干邑总量翻了一倍。

在英国，一些经销商通常会直接从干邑地区进口酒龄较短的桶装干邑，然后将它们放在码头附近的仓库中陈酿。干邑装瓶后，酒标上标示的可能是信誉良好的商号名称，而不是干邑生产商的名字。这些干邑被称为"先抵达"，因其陈酿地点的气候有别于干邑地区，所以味道和香气都与原产地陈酿干邑有细微的差异。

这个时期，白兰地（通常是干邑）在美国已是极受欢迎的鸡尾酒基酒。此时正值贯穿了大半个19世纪的（有史以来第一次）美国鸡尾酒热潮。法国还出现了第一家干邑装瓶厂，这样就可以用玻璃酒瓶取代木桶来装运干邑，让顾客能够一眼认出并欣赏这种独特烈酒所呈现出来的黄褐色。干邑就这样一路发展得顺风顺水，直到19世纪末，所有的一切轰然垮塌。

在整个美国，尤其是南方，干邑向来是有钱人酷爱的烈酒。但是南北战争（1861—1865）之后，南方经济大半被毁，干邑市场因此而凋敝。美国人也开始生产自己的威士忌，并逐渐爱上了波旁威士忌和黑麦威士忌等本土烈酒。

此外，便宜的朗姆酒那时在美国也能买得到了。到了19世纪末，这些烈酒便开始取代干邑在美国的地位——这一点不仅体现在鸡尾酒领域，也反映在美国人的日常生活中。

大约在同一时期，确切地说是1872年，一场灾害重创了干邑的葡萄园。与此同时，因法国在普法战争中输给了普鲁士，拿破仑三世为筹措战争赔款，便新增了一项针对葡萄酒和烈酒的税收，这项额外的税负令法国国内的干邑消费量急剧下降。

干邑葡萄园遭受的病害是一种缓慢蔓延的瘟疫，它用了整整20年来毁掉这些葡萄园。凶手的名字叫作葡萄根瘤蚜，从19世纪末到20世纪初，这种专门破坏葡萄藤的寄生虫肆虐整个欧洲，给一座座葡萄园带来灭顶之灾。在法国，没有一株葡萄能幸免于难，一旦染病即治愈无望，也没有什么办法能消灭这种贪婪的害虫。在干邑地区，根瘤蚜在几十年的时间里一直缓慢传播，最后，到了19世纪90年代早期，葡萄园的面积已大幅缩减，仅余方寸之地。

真正的干邑是举世无双的，因为不管是原料葡萄和当地气候，还是蒸馏师、陈酿师以及市场推广人员的经验，均为无法复制的独门秘籍。但是令干邑生产商倍感苦恼的是，他们发现，由于这种高品质烈酒需求旺盛，利益驱使之下，世界各地许多国家的人也纷纷开始生产白兰地，其中有不少人生产的是次等烈酒，却给它们贴上"干邑"的酒标。例如，19世纪70年代，亚美尼亚和格鲁吉亚的生产商都从干邑地区进口蒸馏器，学习相关知识，为国内市场生产优质"干邑"。19世纪80年代，意大利和希腊生产商开始广泛销售自产白兰地。他们使用的是其他品种的葡萄，出产的产品或许质量上乘（也可能质量堪忧），但为了让自家的烈酒更好卖，都无一例外僭用了"干邑"这个名号。因此，干邑人除了要想办法重建葡萄园外，还要被迫学会如何在世界市场上同其他白兰地竞争。他们不得不走上一条艰难的维权之路，为了捍卫干邑在全球独一无二的名称和身份而向外国生产商宣战。

在19世纪末期的干邑地区，既然找不到长期有效的方法来对付葡萄根瘤蚜，那么就只剩下一种选择：开始在葡萄园中全面改种抗根瘤蚜虫害的葡萄。这是人们用尽千方百计，逐一尝试过宗教疗法、化学疗法和水疗法（淹没葡萄园）之后，终于发掘出来的"良药"。

同法国其他地区一样，干邑也栽种从美国引进的砧木①，然后将欧洲葡萄嫁接上去。在波尔多和勃艮第，用这种方法嫁接梅鹿辄、赤霞珠和黑皮诺等葡萄品种收效甚佳。起初，干邑地区的葡萄种植者嫁接的是几个世纪以来一直用于酿造干邑的白葡萄品种——白福尔和鸽笼白。但是他们发现，白福尔嫁接美国砧木的效果并不理想，于是便尝试将各种葡萄藤同本地的另一种白葡萄品种——白玉霓（又名特雷比奥罗）嫁接起来。同白福尔相比，白玉霓酿就的

---

① 嫁接繁殖时承受接穗的植株。砧木可以是整株果树，也可以是树体的根段或枝段，起固定、支撑接穗的作用，并与接穗愈合后形成植株，继续生长、结果。砧木是果树嫁接苗的基础。

葡萄酒口味稍有不同，也许还要更平淡无奇一些，但用它蒸馏出来的干邑却是美妙绝伦——无论是否加入鸽笼白。最后，干邑地区的葡萄种植和干邑酿制产业终于得以继续发展。

过冬的葡萄藤被修剪整齐，静待下一个生长季的到来。

雅文邑及其光辉历史

今天，位于法国西南部中心地带的雅文邑地区给人一种时间凝滞的感觉，那里的田园风光一如数百年前，甚少变化，而你会觉得这似乎再自然不过。雅文邑是最古老的白兰地产区，比干邑还早上几百年。就在不久前的2010年，名为雅文邑的蒸馏烈酒迎来了700岁"生日"。蒸馏法在雅文邑地区的发展为何如此之早？为何雅文邑白兰地的名声没有干邑那么响亮？根本原因就在于该地区所处的地理位置。

大多数雅文邑酒庄都把生产总部设在乡村，那里有起伏的山峦、小型农场和袖珍牧场。它们可能坐落在村庄附近，或者在村庄外的自家葡萄园内；也有一些位于市镇中心，比如杜培龙雅文邑酒窖就将生产总部建在历史名镇孔东，至今已有100余年。那么，从14世纪雅文邑诞生直到现在，

在雅文邑迷人风光的映衬下，塔西克酒庄里修剪整齐的葡萄植株随山势起伏而绵延铺展。

这期间究竟发生过什么？雅文邑最初又是如何诞生的？

如第一章所述，中世纪早期，蒸馏技术从中东经由伊比利亚半岛传入法国南部。当时，西方世界正在经历一场缓慢的变迁，曾经一统天下的魔法和神秘主义日渐式微，科学开始逐渐占据上风。而医学则横跨了这两个领域，医生还会将一种名为白兰地的葡萄酒蒸馏液开进处方。

第一个将雅文邑蒸馏酒的保健功能编撰成册的是方济各会修士维塔尔·迪富尔（1260—1327）。1310 年，迪富尔写了一部医学专著，论及产自雅文邑的蒸馏烈酒对健康的益处，雅文邑白兰地由此而声名鹊起。他称这种烈酒为"燃烧之水"，人们认为这就是今天雅文邑的直系祖先。

迪富尔的这部著作极为重要，以至于在长达几个世纪的时间里，人们一直以手抄笔录的形式将其代代传承。印

自 20 世纪初朗巴德酒庄出产的白兰地闻名于世以来，酒庄的这座建筑就成了雅文邑乡村地区的一处地标。

今天，这座历史建筑物的彩色外墙为朗巴德酒庄四周的田园风光更添一抹亮色。

刷机发明后，它的读者面进一步扩大。梵蒂冈档案馆如今就收藏了一本1531年出版的迪富尔著作。在书中，迪富尔罗列了饮用由雅文邑"燃烧之水"所制饮品的42个益处。它的神奇疗效包括愈合创伤和溃疡、恢复记忆、治愈肢体瘫痪，以及为胆小之人注入勇气。

接下来的几个世纪里，雅文邑白兰地的产量持续提升，在法国加斯科涅地区的这片区域内满足了当地人对其日益高涨的需求，但此地深处内陆，雅文邑白兰地的发展因此而受到严重影响。实际上，由于没有便于商业运输的河流，雅文邑地区与外界是相对隔绝的。因此，雅文邑或许称得上是法国最早的上品白兰地，但这种烈酒的出口一直困难重重，故此它在全球的名气始终逊于干邑，直到不久前才有所改观。

如今的雅文邑还有一点与干邑大为不同：它是用一次蒸馏工艺制造的。最初，进行一次蒸馏和二次蒸馏的壶式蒸馏器均可用于雅文邑的生产，但在过去的两个世纪，人们采用的一直是雅文邑阿伦比蒸馏器。这是一种体形矮胖的圆柱状一次蒸馏器，于1818年获得专利，经它蒸馏而成的优质白兰地口味与干邑截然不同。

塔西克酒庄使用的传统雅文邑塔式阿伦比蒸馏器，以木柴生火加热。

这种手工一次蒸馏技术还有一个重要优势：蒸馏器体形小巧，便于携带。于是，在雅文邑便会出现这样的情景：蒸馏器被装到马车上，从一处小型葡萄种植园运至另一处，如此一来，就连普通农民都能在自家土地上酿制雅文邑。这种延续至今的习俗和从前雅文邑产量增长之缓慢，使许多生产规模中等的酿酒商得以在自家葡萄园中发展壮大。如今，乡间小巷里的手工酿酒作坊为这一地区增添了无穷魅力。

不过，如今用于酿造雅文邑的葡萄，与19世纪末之前当地选用的酿酒葡萄并不是同一品种。和法国其他地区一样，雅文邑的葡萄园也在根瘤蚜虫灾期间损失惨重。根瘤蚜开始啃食根系、毁掉所有酿酒葡萄植株之前，酸度较高的白福尔是酿制雅文邑的基础，干邑选用的也是这个品种的白葡萄。虫灾向葡萄园袭来时，从事葡萄繁育的法国人弗朗索瓦·巴科将雅文邑和干邑地区的传统白福尔葡萄同美国的诺亚葡萄（其根系具有根瘤蚜抗性）进行杂交，培育出一种抗虫害葡萄。如今许多酿酒商发现，这种巴科22A葡萄虽名不惊人，却能酿出极品雅文邑——尤以种植在最佳产区下雅文邑那片沙砾质土壤中的葡萄为妙。

以葡萄的种植和生产目的为标准，雅文邑被划分为三个产区。大体上自西而东顺序来看，下雅文邑是公认的最优产区，雅文邑-特纳雷泽次之，上雅文邑最末。今天，雅文邑地区选用的酿酒葡萄主要是巴科、鸽笼白、白福尔和白玉霓。允许使用的葡萄品种共十个，其他六个品种名气要小得多，分别是克莱雷、格海斯、白朱朗松、莫札克（白莫札克和桃红莫札克）和梅利耶–圣弗朗索瓦。

在干邑和雅文邑的酿制中，白玉霓葡萄都是关键所在。此图由德洛德雅文邑酒庄提供。

作为一种烈性酒，雅文邑蕴含的果香和花香往往比干邑更为浓郁。另外，若想令雅文邑的各种内在成分在年深日久的陈酿过程中逐渐融合柔化，也需要更长的时间。这就意味着，一瓶口感细腻顺滑的雅文邑，即便只是入门级的，其酒龄通常也会超过顶级的干邑，因而也更加昂贵。

有时候，雅文邑是装在取材于干邑附近森林的利穆赞橡木桶中陈酿的，但如果换用本地加斯科涅森林出产的木桶，它又会别有一番风味，并随着酒龄的增长呈现出越发浓郁的黄金色泽。如今的雅文邑有一套陈酿分级体系，其中包含必须严格执行的各种规定。不过，法国雅文邑行业管理局（简称 BNIA，成立于 1941 年）正在简化雅文邑酒标中使用的描述性术语。管理局提倡设立以下标准：入门级，标注为 VS[1]或三星，规定陈酿时间 1 年以上；中档型，即 VSOP[2]，必须陈酿 4 年以上；Hors d'âge[3]代表真正陈酿10 年以上的雅文邑，酒标上也会注明酒龄（例如 10 年、15年、25 年）；而年份雅文邑必须至少陈酿 10 年，并在酒标上注明葡萄收获年份。管理局还希望在雅文邑地区逐步取消 XO[4]、Vieux[5]和拿破仑等干邑分级标准。

装瓶后，雅文邑就可供人饮用了。它在酒瓶中不会继续陈化，但是启封后，如果置于阴凉处，则可以保存数星

---

[1] Very Special 的缩写，意为"非常特别"。
[2] Very Superior Old Pale 的缩写，意为"品质极佳、酒龄老、酒色淡"。
[3] 意为"超越年龄"。
[4] Extra Old 的缩写，意为"特陈"。
[5] 意为"老的"。

白兰地生产商会将数十年酒龄的熟成白兰地妥善保存在一间名为"天堂"（原因显而易见）的珍藏室里。图中的珍藏室位于德洛德雅文邑酒庄。

在朗巴德酒庄的"天堂"珍藏室，装在细颈大肚瓶中的熟成雅文邑正等待被纳入特别调配版发售产品。

期乃至数月之久。目前，多达500家生产商和300家合作企业每年共出产约600万瓶雅文邑。在英美上架出售的主要品牌包括斯邑男爵、加思德骑士、达豪思、凯龙大帝、德洛德、热辣、金露、朗巴德、乐圣吉、蒙特伯爵、佩莱奥、杜培龙、圣马和塔西克。

相对而言，雅文邑在美国仍然是一种比较罕见的烈酒（在英国的普及度稍高于美国），但它在中国的人气却节节攀升。过去，在长达数世纪的时间里，这种烈酒的分销一直受地理条件所限，但现代交通的发展已经让其贸易往来畅通无阻。在东亚地区，雅文邑这种名贵烈酒以其浓郁醇厚的酒香满足了当地人的口味偏好，且因历史悠久而地位不俗，同时又以精致夺目的包装呈现于世。短短几年间，雅文邑便在中国一飞冲天。到2012年，中国的雅文邑消费量已经超过了美国。

蒙特伯爵雅文邑酒庄中设计美观的现代桶陈酒窖

在德洛德雅文邑酒庄，风格各异的瓶装雅文邑可以根据色彩鲜艳的手工蜡封瓶口区分开来。

一些完全有资格以自家熟成白兰地为傲的生产商，比如雅文邑的朗巴德酒庄，会自豪地将它们摆在陈列室中供人观赏。

为了遵循传统，一些雅文邑的酒标是手工书写的，然后再手工粘贴到酒瓶上。

在德洛德雅文邑酒庄，雅文邑被装瓶后，每只酒瓶都会手工盖上酒庄的徽章。图中这一瓶很特殊，装在里面的雅文邑已有半个多世纪的历史。

欧洲和高加索地区的著名白兰地

干邑在 19 世纪盛名远播，欧洲及其以东地区的一些国家（尤其是位于高加索山脉地区的亚美尼亚和格鲁吉亚）因此而受到鼓舞，纷纷开始生产白兰地。

19 世纪末，一些酿酒厂可能会借用干邑的名气，以较低的成本酿制白兰地，再以远低于进口干邑的价格将其出售。在许多地区，当地出产的白兰地甚至也被冠上"干邑"的头衔，但经过干邑酿造商的艰难维权，这个名称在世界各地逐渐被更改为"白兰地"。在干邑地区之外，大多数白兰地生产商都是用本地葡萄酿造这种烈酒，但也有一些厂商会从干邑进口蒸馏器、引进技术，甚至连葡萄都要从干邑购入。

例如，干邑白兰地在德国享有盛誉，所以德国当地很多白兰地酒厂使用的酿酒葡萄都有一部分甚至全部是从干邑进口的。即便在今天，德国顶级白兰地生产商阿斯巴斯和杜雅尔丁仍是从夏朗德（干邑）地区进口葡萄，采取同样做法的还有远在亚洲（包括俄罗斯部分地区和印度）的一些白兰地生产商。

在意大利，自 16 世纪起就有人蒸馏白兰地，但国家并没有指定的白兰地产区。意大利最卓越的白兰地公司之一始创于 1820 年左右，当时一个名叫让·布东的干邑人前往艾米利亚–罗马涅，发现当地也种有干邑地区的白玉霓葡萄（在意大利叫特雷比奥罗）。于是他建立了一家名为"乔瓦尼·布东"（后来人们也用这个名字来称呼他）的酒厂，开始生产伟杰罗马白兰地——至今它仍是最畅销的意大利白兰地品牌之一。意大利白兰地酒厂斯托克成立于1884年，一些欧洲最著名的白兰地即产自此处。它在20世纪60和70年代达到鼎盛，此后虽不复往日辉煌，但至今仍在意大利和其他国家极受欢迎。

家喻户晓的斯特拉维基奥·布兰卡是另一个顶级意大利商业白兰地品牌，它的生产商是布兰卡兄弟酒厂；意大利人有个治疗咽喉痛的常用药方，就是在一杯热牛奶中加入少许这种白兰地。这家公司于1888年开始生产白兰地，采用的是一种独特的工艺流程：装瓶之前，将一直在"母桶"中陈酿的现存白兰地取出至多三分之二的量兑入，这样最终调配出来的成品就含有陈酿3至10年不等的白兰地。布兰卡兄弟还生产一种特级蒸馏白兰地，名叫麦格纳马特。其他一些意大利公司也生产面向鉴酒行家的高端产品：葡萄酒和格拉帕酒生产商扎里庄园、比安奇侯爵夫人和乔里，以及安东尼世家葡萄酒庄、贝拉维斯塔起泡酒庄和魄力格拉帕酒庄，都是兼酿白兰地的酒商。

穿过地中海来到希腊，我们会发现最著名的希腊"白兰地"迈夏尔有个颇具妙趣的小番外（许多旅行者在地中海度过热情似火的假期后，都会对这种酒留下深刻印象）。1888年，斯皮罗斯·迈夏尔开始生产与自己同名的烈性酒，随后迈夏尔酒的名声传遍全球。他生产的这种烈酒与当时品质粗劣的本地酒迥然相异，因而受到了市场的欢迎。不过，他还在自己的蒸馏酒中加入了甜麝香葡萄酒和草药，所以严格来说，经典迈夏尔并不是（经典）白兰地。

白兰地是亚美尼亚的重要产业，2007年发行的这枚邮票就是为了纪念亚美尼亚白兰地之父尼古拉·舒斯托夫。

　　继续向东行进，高加索地区的亚美尼亚和格鲁吉亚也出产优质白兰地。这些国家拥有许多产量丰富的白兰地公司，长期以来，它们一直用夏朗德式（干邑式）蒸馏和陈酿法生产优质白兰地，供俄国沙皇享用。但由于所处地理位置遥远，加之"铁幕"的阻隔，它们一直不为西方所知。传说1917年布尔什维克攻陷冬宫后，革命全面暂停了一个星期，在此期间，革命者们将沙皇那些妙不可言的极品白兰地喝了个精光。

就在几十年前，亚美尼亚还是为俄罗斯及其他苏联加盟共和国供应白兰地的主要指定生产国。苏联解体后，分销网络一夜之间消失殆尽，亚美尼亚白兰地的市场也随之崩塌。目前，亚美尼亚的白兰地产业正在重建；从前的市场已经稳定下来，同时生产商也在积极开拓新市场。

如今，亚美尼亚有三家大型白兰地生产公司，还有一些公司正处于私有化的艰难关口。有几家仍然在首都埃里温，其设施通常比较先进；其他公司则位于这个多山国家边缘高海拔的干旱高原上。主要的公司有亚拉拉特、诺伊和普罗尚，还有曾经显赫一时、如今刚刚东山再起的韦迪阿

这些酒瓶和酒桶可以证明，在亚美尼亚的埃里温市，优质白兰地的生产历史已逾一个世纪。

埃里温白兰地公司的酒窖处处散发着明星魅力：一条条过道里满是纪念亚美尼亚名人的各种物件，酒桶上还可以看到社会名流和政治家的签名。

尔科等其他一些生产商。

令人困惑的是，亚拉拉特和诺伊有时都会被称作埃里温白兰地公司。亚拉拉特以《圣经》中诺亚方舟登陆的那座山命名；诺伊则是亚美尼亚语中的诺亚。两家公司都宣称自己于1877年在亚美尼亚开创了白兰地产业。最初，一家白兰地蒸馏厂和若干陈酿酒窖在一处16世纪波斯城堡的遗址之上兴建起来，从城堡居高俯瞰，可以望见通往埃里温的道路。1899年，这家公司被俄国行业大亨和白兰地推广者尼古拉·舒斯托夫收购，此人也被视为亚美尼亚支柱产业——白兰地产业——之父。这个产业在亚美尼亚历史上发挥过至关重要的作用，为此，亚美尼亚还在2007年发行过一枚印有舒斯托夫照片的邮票。

1912 年，舒斯托夫的公司成为沙皇尼古拉二世宫廷的官方供应商。苏联时期，他的白兰地在俄罗斯和其他苏联加盟共和国一直极为抢手。据说，1945 年雅尔塔会议期间，斯大林还让丘吉尔初次品鉴了亚美尼亚白兰地。丘吉尔显然爱极了这种酒，据传斯大林在此后余生的每一年都会给他寄去一整箱。（在接下来介绍格鲁吉亚白兰地的部分，还会讲述另一个关于丘吉尔的故事。）

埃里温白兰地公司在现代化会议室里举行以餐后甜点搭配亚拉拉特白兰地的品酒会。

苏联时期，舒斯托夫的公司更名为埃里温，并于 1950 年将生产总部和陈酿酒窖迁至该市一处现代化的新厂区。1998 年，公司被国际大型葡萄酒和烈酒集团保乐力加收购，如今名叫亚拉拉特，采用现代化生产设施，以干邑式蒸馏法生产其著名的白兰地。目前，亚拉拉特从 5000 个葡萄种植园收购葡萄，每年产量达 550 万瓶，其中有 92% 出口到俄罗斯和波罗的海国家。

亚拉拉特的酒窖里还存有一只"和平桶"，它将 1994 年亚美尼亚和阿塞拜疆出产的年份烈酒融合，象征双方争夺的纳戈尔诺-卡拉巴赫地区在这一年实现停火。这个酒桶于 2001 年举行封盖仪式，等待该地区正式签署和平协议之时再开启。

埃里温老字号诺伊白兰地公司现在的酒标，上面印有纪念诺亚方舟停在亚美尼亚亚拉拉特山的图案。图中的酒瓶里分别装着陈酿 10 年和 20 年的白兰地。

普罗尚白兰地公司及其富有建筑美感的陈酿酒窖代表了后苏联时期亚美尼亚的现代化国际白兰地产业。

苏联解体后，私人投资者凑足资金，于2002年在舒斯托夫工厂原址重启生产。他们将这家风投企业命名为诺伊，公司徽标以诺亚方舟为图案，上面还注有"1877"这一年份。显然，先前的公司并未将其全部窖藏都转移到新址，因为如今，在地下深处的诺伊酒窖里，仍可找到有近百年历史的白兰地。亚美尼亚的白兰地产业兜兜转转又回到原点：2011年，诺伊新开发了一系列专供克里姆林宫的白兰地。

据公司所有者说，普罗尚是一个古老而高贵的名字，公司总部所在的埃里温郊外村庄也以此为名。曾有一家"普罗尚白兰地公司"成立于1887年，但现在这家公司并不是它的直系后裔。苏联时期公司尚未创立，尽管其部分生产设备的历史可以追溯到那个年代，但如今从意大利引进的新机器已令公司实力大增。2012年，一座用玻璃和大理石建成的新办公楼正式启用，为原有的旧厂区更增一抹亮色。普罗尚是一家现代感十足的公司，目前它为 500 家欧洲超市生产品牌酒，其产品还销往俄罗斯、德国、波罗的海国家、韩国和中国。不过，公司名气的来源之一却是其精致华美的传统酒瓶，有玫瑰、船只、宝剑和飞龙等各种迷人的造型。普罗尚的市场江山足有 20% 是靠这些"纪念品"酒瓶打下的。

韦迪阿尔科的境况与普罗尚的欣欣向荣形成了鲜明对比。这家公司位于一处苏联时期的乡村老厂，距离埃里温只有几小时车程。苏联解体后，1996 年，一群急于改善自身经济状况的工人冒险接管了它。原厂创立于 1956 年，厂内设施至今仍留有 20 世纪中叶苏联时期的印记。2000 年，公司重新开始在这里蒸馏白兰地，还收购了一些陈年烈酒。目前，公司用塔式蒸馏器进行间歇蒸馏，但还希望能添置一套从前使用过的那种二次蒸馏系统——尽管最近修葺屋顶又成了当务之急。如今，韦迪阿尔科白兰地的市场需求刚刚开始回暖，其产品销往海外，主要面向俄罗斯市场。

在亚美尼亚文化中，白兰地被视为一种用来搭配甜点的饮品，可佐以巧克力，或柑橘和苹果，或成熟的时令鲜桃。就像在西方许多地区一样，亚美尼亚人也有在晚餐后边喝白兰地边抽雪茄的传统。根据礼节，如果主人将白兰地盛在专用的窄口杯中递给客人，那么客人应当能够将酒杯横放而不会洒出一滴酒液——这意味着主人斟出的恰好是可供啜饮的量，即 50 毫升左右（约 1.7 液量盎司）。

目前，用于生产亚美尼亚白兰地的葡萄酒可以使用13个品种的葡萄（主要是白葡萄）来酿制：阿扎特尼、班安提斯、奇拉尔、加朗玛克、卡赫特、坎根、拉尔瓦里、马西斯、梅格勒布杰尔、慕斯卡里、白羽、梵和沃士奇。

苏联时期，亚美尼亚的邻国格鲁吉亚就以葡萄酒而闻名于世，不过此地自19世纪后期以来也生产白兰地。1884年，格鲁吉亚人大卫·萨拉吉什维利创立萨拉吉利，130年来它一直是格鲁吉亚最重要的白兰地品牌。萨拉吉什维利本人曾在德国研习化学，在干邑学习蒸馏技术，但父亲去世后他不得不返回故土。在国内，他奔走于琴纳里、果露丽-莫茨瓦尼、卡胡里-莫茨瓦尼、尔卡齐杰里、吉斯卡和索丽科

傲然成立于1956年的韦迪阿尔科公司自苏联解体以来一直举步维艰。如今它由原厂工人拥有和管理，用苏联时期遗留下来的设备生产葡萄酒和白兰地。亚美尼亚是苏联全盛时期的主要白兰地供应国。

几尊雕塑装饰着韦迪阿尔科公司的地面和台阶，气势宏伟的正门上分别以俄语和亚美尼亚语标注着公司名称。

里等地，从 500 个格鲁吉亚本土品种中，筛选出同干邑地区使用的葡萄具有相似特征的品种。对于格鲁吉亚白兰地生产商来说，从全国各地采购葡萄已经成了一种习惯。

萨拉吉什维利还从法国引进了另外两大要素：一是一只来自干邑的铜质阿伦比蒸馏器，二是该地区最古老的家族制造企业卡慕干邑方面的人脉。双方业务往来持续了数十年之久，但在苏联时期被切断。后来，公司接洽卡慕现任掌门人的父亲，将合作关系又重新接续起来。

苏联时期，公司更名为第比利斯白兰地公司，其产品经常被当局征用。不过，这家酒厂仍然设法留存了一些具有历史意义的桶陈酒，其中有几桶还是大卫·萨拉吉什维利在 1893 年和 1905 年亲自酿制的。如今这些年代久远的酒

一只古老的格鲁吉亚酒具，由三只酒杯彼此相连铸成，可能是在起源于数千年前的某种饮酒仪式中使用。

格鲁吉亚第比利斯国家博物馆收藏着一只数千年前的角状酒具，以整块石头凿刻而成，表面有精美雕饰。

同干邑地区一样，格鲁吉亚的萨拉吉利"干邑"也是用葡萄酒蒸馏而成，从19世纪末开始屡获奖章奖项。

桶都储藏在酒窖里，公司会从中取出少许陈年佳酿，兑入其生产的顶级白兰地中。

萨拉吉利公司的总部位于第比利斯，是1954年兴建的一座花园式酒庄。1994年公司实行私有化之后，采用铜质蒸馏器二次蒸馏的工艺，继续生产一系列陈酿白兰地。熟谙公司历史、对产品具有前瞻眼光的首席技术专家大卫·阿布齐亚尼泽继承了萨拉吉利白兰地的传统。不过，尽管其他国家非常流行用白兰地调制鸡尾酒，但他丝毫不为所动，无意将自己的白兰地派作此用。

阿布齐亚尼泽讲述了一个（似曾相识的）故事，是他从上一任首席技术专家那里听来的：1945年雅尔塔会议期间，斯大林送给温斯顿·丘吉尔一些萨拉吉利白兰地，丘吉尔认为它的品质丝毫不输干邑，于是把它选为会场上的最佳白兰地。

现如今，还有几家格鲁吉亚公司借用格鲁吉亚白兰地（尤其是萨拉吉利）的地位生产白兰地。它们可能会选用产自格鲁吉亚任意地区的葡萄，将产品装瓶销往熟知格鲁吉亚白兰地大名的国家。第比利斯马拉尼葡萄酒公司就是其中之一，尽管目前它最著名的产品仍是葡萄酒。另一家是卡赫季传统酿酒公司，简称KTW。这是一家年轻的公司，员工也都是一群朝气蓬勃、精力充沛的年轻人。公司成立于2001年，其生产的中等价位葡萄酒和白兰地在东欧和波罗的海国家大获成功。KTW凭借怀旧风格的包装为自己打开了一片缝隙市场，它的许多款白兰地，无论是包装瓶还

萨拉吉利在苏联时期生产的白兰地在品质上或许无法同之前和之后相比，但公司仍设法保住了自己的地位，还留下了位于第比利斯的这片宽敞园区，内有一座座爬满藤蔓的建筑和公司创始人的雕像。

萨拉吉利"干邑"曾专供俄国皇室享用；这则广告以崎岖山峦和一只成功登顶的山羊为象征，诠释了品牌中蕴含的力量。

是随身瓶的造型，在外观上都给人以手工打造的感觉。

虽然这些格鲁吉亚和亚美尼亚白兰地大多尚未进入英美市场，但更多产品上架应该只是时间问题——不过，它们抵达的第一站可能是热切的亚洲市场。

这家为自己取名"卡赫季传统"的年轻公司，当初创办时就是以传统主义者为目标顾客的。在其位于第比利斯的办公室中，公司将自己生产的白兰地和所获奖项陈列展示。

西班牙和拉美的杰出白兰地

虽然人们通常认为白兰地属于法国酒，但在西班牙和拉美文化中，白兰地也拥有可追溯到数世纪前的悠久传统。在本章中，我们会了解到伊比利亚半岛上的白兰地及其传承，以及秘鲁在此领域做出的重大贡献。

西班牙是旧大陆的一种名贵白兰地——赫雷斯白兰地的故乡，而这种酒如今的名声本应更加响亮才是。它出产于赫雷斯地区，这里也是雪利酒的发源地，位于西班牙西南部，距离直布罗陀海峡附近的大西洋南部海岸不远。赫雷斯白兰地能带给人独一无二的感官享受，这不仅同赫雷斯所处的地理位置有关，也得益于此地独特的白兰地陈酿系统。赫雷斯白兰地和雪利酒采用的是同一套陈酿法，即索雷拉[①]系统（见下文的讨论）。这种酿酒方法会赋予白兰地独特的味道和香气：隐隐约约的海盐、香草和烤焦糖味，还有角豆和咖啡香，最深层的香气则让人联想起成熟橡木和酵母的味道。

公元前 700 年至公元前 500 年左右，腓尼基人在西班牙海域四处航行，自那时起，赫雷斯地区就发展起葡萄酒和其他产品的贸易。罗马帝国时期及帝国衰亡后，这里的葡萄酒也一直远销海外，直到公元 711 年至 1492 年摩尔人占领伊比利亚半岛期间才被迫中止。摩尔人不饮酒，但他们利用现成葡萄园中的葡萄生产酒精蒸馏液，用于制造药物、化妆品和香水。

[①] 原文 solera 在西班牙语中意为"在地面上"，在酿酒术语中指的是一套酒桶设备中位于最下方贴近地面的那一层。

西班牙赫雷斯白兰地在橡木桶中陈酿，橡木的颜色沁入清澈透明的酒液，使其渐次转变为黄色、金色、琥珀色和桃木棕色。

后来，摩尔人被驱赶回南方，最终又被逐出伊比利亚半岛，但他们的蒸馏技术却保留了下来，当地医药专家和炼金术大师就运用这种工艺自制"生命之水"和"精魂之水"。实际上，白兰地在西班牙语中至今仍被通称为 aguardiente[①]（来自 aqua ardens 一词，即中世纪的"燃烧之水"）。

最早提及赫雷斯白兰地的史料是 1580 年一份关于烈酒税的记载，但白兰地在此地的生产史可能要追溯到更早的时间。正如在法国一样，在西班牙的这一地区，开白兰地制造之先河的也是荷兰人，时间是在 16 世纪末。今天，用壶式蒸馏器蒸馏出来用于陈酿白兰地的清澈烈酒仍被称为"荷兰酒"。在赫雷斯-德拉弗龙特拉的许多大街、人行道上和庭园中，都可以看到一排排的小圆石，它们于数世纪前从荷兰远道而来，当时的船只在赫雷斯装载蒸馏烈酒时，就是用这些石头做压舱石的。

---

① 西班牙语中，agua ardiente 直译为"燃烧之水"；aguardiente 为合成词，意为蒸馏酒。

在 18 世纪，这种西班牙白兰地一直都是未经陈酿就直接装船运走的；事实上，根据行会（葡萄种植者联盟）的规定，"荷兰酒"每年都要被运往海外，好让种植园和酿酒厂在下一个收获季到来前清空自己的酒窖，并迅速回笼货款。

相传在 19 世纪初，有一批蒸馏酒被装进用过的雪利酒桶里准备装船运走，但是船只离港出发后，才有人发现这些木桶被遗忘在了原地。酿酒商品尝了这种烈酒，发现经过木桶储藏之后，酒的味道竟然更加醇美，这就是赫雷斯桶陈白兰地的起源。当时的酿酒商就是如今赫赫有名的派卓多美公司，而那一年是 1818 年。

19 世纪，赫雷斯地区的白兰地生产和贸易一直稳步增长。这种白兰地最初用当地的帕洛米诺葡萄酿造（这种葡萄也用于酿制雪利酒），但 19 世纪后期的白兰地热潮迫使生产商将目光投向更远的地方。他们发现，西班牙中部卡斯蒂利亚-拉曼查地区栽种的艾伦葡萄恰好具备所需的特质。

赫雷斯白兰地所用的大部分葡萄如今仍然是在卡斯蒂利亚-拉曼查种植的。这种葡萄在拉曼查地区的托梅略索市被酿成葡萄酒，许多白兰地生产商还在此地设有自己的蒸馏厂。不过，白兰地的调配和陈酿一直都是在赫雷斯地区进行，确切地说，是以赫雷斯、圣玛丽亚港和桑卢卡尔-德巴拉梅达这几个城镇为界圈出的沿海"雪利金三角"地区。部分白兰地也会在赫雷斯蒸馏。

壶式和塔式蒸馏器都用于赫雷斯白兰地的蒸馏。如果

门多萨主教是全球最著名的西班牙白兰地品牌之一。图中此酒正在西班牙赫雷斯的山穗洛梅酒庄桶陈，之后会被装瓶。

严格区分的话，由蒸汽加热的蒸馏器叫作阿伦比蒸馏器，而用木柴加热的则被称为阿尔基塔拉蒸馏器。以这两种蒸馏器蒸馏出的烈酒都被归类为"荷兰葡萄酒"。

赫雷斯的桶陈仓库（酒窖）和办公室一般都是粉刷成白色的土砖结构，楼高数层，楼顶覆以橘红色陶土屋瓦。其中有许多位于赫雷斯市中心，形似微缩版的 19 世纪别墅：围墙后一座座建筑鳞次栉比，深红色九重葛点缀其间，将它们衬托得更加悦目。这里一年四季空气清新宜人，每一处繁忙的营业区都建有以棕色鹅卵石小径连接起来的白色楼宇，中间往往穿插着花园或露天广场。

赫雷斯白兰地的酒桶（以前用于陈酿雪利酒）由美国橡木制成，这是从该地区首度成为大西洋沿岸繁华贸易区时起传承下来的习俗。人们会选择曾经陈酿过不同风格（从甜型到干型）雪利酒的酒桶来塑造这种白兰地的香气和味道。

赫雷斯白兰地全部采用索雷拉法生产，传说这是缘于另一桩幸运的意外事件。1870 年，许多桶白兰地被堆放在一座酒窖的角落里无人认领。1874 年，有人发现了这些酒桶，但一时半会儿没法把这些陈年白兰地全部卖出去，于是他们开始进行调配，将一部分新酿白兰地注入桶中，填补正常蒸发留下的空间。这种方法大获成功，陈年烈酒的质感因此而变得更加丰富细腻，于是索雷拉法从此被沿用下来。

按照索雷拉法，一桶桶白兰地在高顶酒窖中根据酒龄长短逐层叠放。最顶层酒桶里装的是最年轻的白兰地，然

西班牙赫雷斯的一座典型陈酿酒窖

后依次向下，白兰地的酒龄一层比一层长，一直到地面层（索雷拉层）。将一部分熟成白兰地从地面层酒桶中取出以备装瓶后，地面层酒桶中腾出的空间会由上一层酒桶中的酒加满，上一层继续由再上一层加满，依此类推。索雷拉层上方的所有层被称为 criaderas，在西班牙语中，这个迷人的词意为"苗圃"。通常来说，酒窖中至少会垂直叠放三四层陈酿桶，上方留有足够的空间。

在干燥炎热的夏季，人们会在酒窖的红棕色黏土地面上洒水，以维持室内湿度，减少桶内酒液的蒸发量。每年从木桶中蒸发掉的白兰地比例最高可达 7%，因此一些酒窖开始引入温度和湿度控制系统。虽然这种做法可能有助于提高公司利润，但如果没有自然微风从酒窖中吹拂而过，白兰地是否仍能保留当地独有的香气和味道，还有待时间来验证。

在赫雷斯，顶级雪利酒生产商往往也能酿出最优质的白兰地，不过这些白兰地却并不是最畅销的。在特里酒庄及其所产的"百年白兰地"身上就出现了这种反差。特里酒庄由一个爱尔兰家族创立于 19 世纪中叶。20 世纪初，他们在赫雷斯附近的圣玛丽亚镇建起新酒厂，于是便将自己的白兰地命名为"百年"。

随处可见的门多萨主教也是赫雷斯最著名的白兰地品牌之一，诞生于 1887 年，生产商是罗曼彻酒厂。这种酒最初酿制出来是供酒厂老板们自己享用的，但很快就被推向市场，如今已誉满全球。

有的酒窖会购买现成的烈酒进行陈酿，而拥有相关设备的酒厂则是自己制作大部分蒸馏酒，例如150年老厂百艾斯和始创于18世纪的派卓多美（现已并入芬德多酒庄）。间或也会零星出现几家新公司，比如成立于1998年的传统酒庄。它从多种渠道购进窖藏白兰地，之后进一步陈酿，调配装瓶后再贴上自己的酒标。

虽然赫雷斯已有数世纪的白兰地生产史，但直到1987年，赫雷斯白兰地才建立了自己的监管委员会。该委员会规定，根据赫雷斯白兰地的"挥发性成分"递增量和陈酿级别，可以将其划分为三个等级：索雷拉赫雷斯白兰地，平均陈酿期一年半；索雷拉珍藏赫雷斯白兰地，平均陈酿期3年；索雷拉特酿珍藏赫雷斯白兰地，平均陈酿期10年。

19世纪末白兰地迎来繁盛期之后，西班牙其他地区也开始少量生产白兰地。在东北部的加泰罗尼亚大区，人们为了挨过山区的恶劣气候，曾经将白兰地作为必要的热量来源之一。直至20世纪末（也许今天依然如此），这里的工人仍然习惯在清晨上班途中到咖啡馆稍作停留，来一杯加白兰地的咖啡暖身提神。

加泰罗尼亚的白兰地生产商中有一家叫马斯卡罗，成立于第二次世界大战末期。历史上，许多加泰罗尼亚人曾在西班牙殖民地酿造朗姆酒，然后再将其产品出口回西班牙。但在西班牙内战期间以及随后因二战而导致的贸易中断期，西班牙不再进口烈酒。于是，加泰罗尼亚人只好开始利用手头的原材料（葡萄酒）在本地自行蒸馏烈酒。

由于居住地毗邻法国边境，加泰罗尼亚人便选用夏朗德式或干邑式二次蒸馏法，比如纳西索·马斯卡罗（其父是酒商兼酿酒师）就是这样做的。西班牙起泡酒卡瓦同样出产于这一地区，而且用于酿制卡瓦酒的葡萄非常适合蒸馏。帕雷亚达葡萄酸度高、香气清雅细腻，是酿制许多加泰罗尼亚白兰地的主要品种；除此之外也会用到马家婆葡萄和沙雷洛葡萄。

同样位于伊比利亚半岛的葡萄牙也生产一些白兰地，尤其是在洛里尼扬市。自18世纪以来，此地就有蒸馏白兰地的传统，但直到20年前，这个历史悠久的白兰地产区才获得DOC（原产地名称保护，英文缩写为PDO）地位。洛里尼扬位于里斯本大区北部的一处葡萄酒酿酒区，人们用传统葡萄牙语中的"燃烧之水"一词来称呼这里生产的白兰地。

在大西洋彼岸，西班牙前殖民地之一墨西哥以其昔日宗主国的方法为蓝本生产白兰地。墨西哥白兰地也采用索雷拉法生产，不过在世人看来，其品质通常不及西班牙白兰地。21世纪之前，墨西哥的酿酒葡萄几乎全部用于蒸馏白兰地。该国最著名的白兰地品牌是总统，由赫雷斯的派卓多美酒庄在墨西哥生产，不仅畅销国内市场，还出口到全球许多国家。

19世纪初之前，统治菲律宾的殖民政府总部一直设在墨西哥而不是西班牙，因此菲律宾文化深受墨西哥影响。西班牙和墨西哥的白兰地都出口到菲律宾，而且西班牙的芬德多白兰地在菲律宾的地位非常重要。另外两个在菲律宾热卖的品牌是本地出产的皇胜和赫内罗素。

另一方面，16世纪时白兰地也在南美洲同步发展起来，尤其是在皮斯科白兰地的故乡秘鲁。皮斯科的确是白兰地的一种，虽然很多人并未意识到这一点。与其他白兰地不同的是，它是一种白色（透明）烈酒，而且蒸馏后不加水。换言之，烈酒从蒸馏器中馏出时，就必须达到可供装瓶的理想酒精度，即38%至43%之间。而赫雷斯白兰地、干邑和大部分其他白兰地都是先被蒸馏至较高的酒精度，然后在装瓶前的几个月里逐渐兑水稀释。

早在17世纪初，皮斯科就在秘鲁南部沿海地区诞生了，而当地种植葡萄的历史始于16世纪中叶。这种酒得名于港口城市皮斯科，大部分皮斯科白兰地正是在这里装船出海。当时，葡萄酒是西班牙宗教和文化中的重要元素，因此西班牙探险家和移民从加那利群岛及西班牙本土带来葡萄，将它们移植到秘鲁南部海岸附近。结果，葡萄在这一地区竟然大获丰收，以至于没过多久，葡萄酒就开始从秘鲁出口到西班牙。然而，西班牙酿酒商强烈抵制来自殖民地进口产品的竞争，因此一项禁止秘鲁葡萄酒进口的法律于1641年颁布实施。于是，秘鲁酿酒商便开始将葡萄酒蒸馏成白兰地（烧制葡萄酒）后再出口到西班牙，结果恰好投当地市场所好。

芬达多白兰地问世于100多年前，当时它就装在这样的瓶子里出售，外包装纸上印有白兰地陈酿地点的全称：西班牙赫雷斯-德拉弗龙特拉。

最初，秘鲁白兰地可以用两种蒸馏器中的任意一种蒸馏：一种是法尔卡蒸馏器，它的构造类似于早期摩尔式或中世纪时期的某种蒸馏器——一个简单的加热容器，外加一根用于冷凝蒸汽的长管；另一种是秘鲁版阿伦比蒸馏器，相对来说构造要复杂一些。同欧洲的阿伦比蒸馏器一样，它也装有长长的螺旋形铜质管颈。在后一种工艺流程中，酒头和酒尾被丢弃，皮斯科蒸汽形成的纯净酒心则被保留下来，经冷却后倒入陶土罐中等待被装船运走。

虽然17和18世纪旧金山地区一直在进口皮斯科，但到了19世纪中叶，当第一批淘金者抵达加利福尼亚时，皮斯科的需求出现了暴涨。在 19 世纪余下来的时间里，皮斯科的名气有增无减。20 世纪初，旧金山掀起一股皮斯科热潮：它让人怎么喝也喝不够。人们常常把它做成皮斯科潘趣酒饮用，据说这种酒也是在旧金山发明的。据当时的报纸报道，这种加了柠檬、白糖和菠萝调制而成的皮斯科潘趣酒还穿越加州传入了内华达州。

20 世纪早期，尽管皮斯科和皮斯科潘趣酒在西方广受欢迎，但禁酒令实际上摧毁了美国的皮斯科市场。之后，皮斯科的名声一落千丈，沦为一种廉价的粗制烈酒——在人们印象中，这种酒多半是老西部片中加州酒吧的标配。

但皮斯科并未就此消亡——实际上，新千年到来之际，它已蓄势待发，准备重拾辉煌。正宗皮斯科的魅力被前往秘鲁的游客重新发掘出来，酒厂靠升级版皮斯科再次打开市场。尽管皮斯科的酿制史已有数世纪之久，但直到 1999 年，秘鲁生产商才制定了自己的原产地名称保护法，对产区、葡萄的品种和质量以及烈酒的蒸馏和陈酿标准作出规定。

进入 21 世纪才仅仅十年，皮斯科的人气便开始再度飙升，尤其是在当时的调酒运动中。如今，这种烈酒有两种制法——纯皮斯科和皮斯科混酿。前者由单一品种的葡萄酿制而成，后者至少会用到两种葡萄，通常芳香型和非芳香型葡萄均有使用。

另外，用于蒸馏皮斯科的葡萄酒可以是完全发酵型，也可以是部分发酵型（仍然保留些许甜味），有时还会将二者混合蒸馏。可以酿造皮斯科的葡萄共有八种：用于混酿皮斯科的芳香型葡萄品种有意大利、特浓情、麝香和阿比洛，而非芳香型皮斯科（更为典型）一般是由酷斑妲、莫雅尔、黑克里奥拉和乌比纳葡萄制成。到目前为止，酷斑妲葡萄是酿制秘鲁皮斯科的最主要品种。

在数世纪的时间里，秘鲁几乎成了皮斯科的代名词，但今天它已不是这种烈酒的唯一生产国：智利也开始行动起来，力图争夺全球皮斯科市场。实际上，智利生产皮斯科式白兰地的传统由来已久，尽管就在 10 年前，智利人为了喝到品质更好的皮斯科，仍是会将目光投向秘鲁。但如今，秘鲁皮斯科一家独大的局面已经一去不返了。

然而，智利的皮斯科无论是香气还是味道都与秘鲁版颇为不同——智利皮斯科口感更柔和，香气更浓郁。目前，智利有多家生产商采用干邑式二次蒸馏器，选用芳香型的麝香葡萄生产出品质上佳的皮斯科。其中最著名的公司有卡帕（出自酿酒世家马尼埃-拉博丝特）和 ABA 等。这些生产商选用的都是优质酿酒葡萄，最终的成品酒因此而口味非凡。

尽管秘鲁仍然是皮斯科产业的龙头老大，但实际上智利早在 1936 年就有了自己的皮斯科原产地名称：当时葡萄种植区埃尔基省的一个小镇更名为"皮斯科埃尔基"。目前，智利生产商还在尝试干邑酿制中的另一个要素——桶陈。

在拉丁美洲的其他地区，优质白兰地的生产并没有一定之规。例如，阿根廷品牌拉美福干邑就是在法国干邑生产商指导下创立的，其白兰地也是采用干邑风格酿制。而玻利维亚则生产一种别具风格的本国特色白兰地，名为辛佳尼。这种独特的烈酒主要用高海拔地区种植的麝香葡萄蒸馏而成。

在全面梳理了这些西语世界的白兰地之后，我们将继续踏上旅途，去探索世界其他地区的白兰地。但在此之前，让我们暂且先放松一下——不过，究竟是来一杯赫雷斯白兰地还是皮斯科酸鸡尾酒，倒真是令人难以抉择。

澳大利亚和南非

大英帝国的居民都是白兰地的忠实拥趸，这一口味上的偏好，有人是从自己祖国带过来的，有人则是到了国外才培养起来。如前文所述，白兰地亦被视为一种家庭常备药，而且这种情况一直延续到20世纪。

同欧洲远隔重洋的英联邦国家（尤其是南非和澳大利亚）自己生产白兰地供国内消费。例如，在澳大利亚，知名品牌塔南达酒庄在一个多世纪之前就已得到广泛认可。这种白兰地产自该国最早的葡萄种植区之一——澳大利亚南部的巴罗萨谷。1890年，塔南达酒庄在此地建立，而这时距离巴罗萨谷开始种植葡萄已经过去了几十年。不过，凭借酒庄那宏伟壮丽的庄园和公司精妙的销售策略，塔南达白兰地稳获"英联邦医用白兰地"的头衔。

人们笃信白兰地包治百病，能让一切难题迎刃而解。

澳大利亚的塔南达酒庄从19世纪末期开始生产葡萄酒和白兰地，时至今日，塔南达酒庄白兰地仍是极受欢迎的品牌。

事实上，白兰地还曾被奉为挽救运动生涯的功臣：1896年，《珀斯每日新闻》在头条报道中说，著名板球击球手弗兰克·艾尔代尔"因白兰地和苏打水而扭转败局"。

19世纪中叶，澳大利亚巴罗萨谷种植的葡萄既用于酿制葡萄酒，也用于蒸馏白兰地。从18世纪中期开始，甜如糖浆的强化葡萄酒、雪利酒式葡萄酒和口味醇厚的白兰地便风靡澳大利亚。然而遗憾的是，澳大利亚人对白兰地的认知止步于旧时代——如今对白兰地感兴趣的年轻人已寥寥无几。据澳大利亚白兰地生产商说，人们认为白兰地这种烈酒还算过得去，不过说来也怪，它的受众面非常狭窄，主要是40岁以上的女性。

虽然从1855年起，安戈瓦家族就已经在澳大利亚酿造葡萄酒，但直到1910年，他们才开始种植专用于制造白兰地的葡萄。1925年，卡尔·安戈瓦在澳大利亚开办了第一家工业酿酒厂。他遵循法国和西班牙的模式，选用中性香型但可以在酸度较高时采收的葡萄品种：鲜食葡萄苏丹娜，以及法国的鸽笼白和西班牙的帕洛米诺。大多数澳大利亚人喝惯了口味醇厚甘美的葡萄酒和烈酒，但安戈瓦白兰地的风格与之稍有不同。人们觉得它更接近于那种味道清爽的干邑式白兰地，对它很是喜爱——至今依然如此。安戈瓦旗下品牌圣格诗在其故乡南澳州占领了70%以上的市场，在全国白兰地市场占有的份额为40%。据说，如今仍使用夏朗德壶式蒸馏器且以二次蒸馏法生产白兰地的公司，安戈瓦是独一家。

除了安戈瓦和塔南达酒庄，在澳大利亚最走俏的白兰地还有夏迪黑瓶和沃尔沃斯超市经销的法国品牌拿破仑1875；人头马出产的法国干邑口碑也不错。在相当一段时间里，澳大利亚的白兰地消费始终不温不火，白兰地鸡尾酒运动也未见兴起之势。不过最近又出现了一个新市场——中国。虽然中国人对欧洲白兰地非常迷恋，但他们也在寻求其他渠道来满足自己对高端白兰地的渴求。于是，他们突然对安戈瓦 XO 等酒款起了兴趣，而这可能会掀起一波新潮流。

印度与澳大利亚正相反，在这里，白兰地的主要消费者是男性。他们饮用的高档烈酒通常进口自干邑地区，但

1925 年，安戈瓦家族酒庄开始在澳大利亚生产著名的圣格诗白兰地。

中低档烈酒一般都是在印度生产的。印度本土并不盛产葡萄，葡萄酒的产量也比较少，因此印度酒厂往往会用糖类制品蒸馏出烈酒——所以严格来说，这些烈酒其实应该叫朗姆酒，而不是白兰地或威士忌。但它们都是有色酒，有时也会经过陈酿或调味，或两者兼有，再贴上白兰地或威士忌的酒标卖给早已喝惯此类烈酒的受众。在马来西亚和菲律宾等其他一些亚洲国家，白兰地的消费量也很高。但它们本国出产的所谓"白兰地"，采用的多半也是同朗姆酒生产工艺类似的流程。

相比之下，另一个前英联邦国家南非①由于最早是荷兰殖民地（可追溯到 1652 年），因此极为偏爱干邑风格的白兰地。1659 年之前，早期殖民者就已经在南非种植葡萄。关于白兰地在南非的起源，有一种流传甚广的说法：1672年 5 月 19 日，一艘名为德皮尔的荷兰船在近海处停泊，南非历史上第一次白兰地蒸馏就此开始。如前文所述，17 世纪时，荷兰人曾在法国多个地区将葡萄酒蒸馏成白兰地，所以他们在这方面所拥有的知识和设备使得他们无论在哪里，只要能找到适宜的葡萄品种，就可以动手蒸馏白兰地。

由于可供蒸馏的白葡萄数量充足，南非的白兰地产业与葡萄酒产业得以同步发展。在这里，用于酿制白兰地的主要葡萄品种是白诗南和鸽笼白（在南非，鸽笼白通常拼作 Colombar 而不是 Colombard，少了最后一个字母 d）。

---

① 此处原文疑有误，经查证，南非于 1961 年成为共和国时退出英联邦，后于 1994 年重新加入英联邦。

在长达数世纪的时间里，南非生产的白兰地几乎全部供国内消费。诞生于1845年的范瑞斯是早期一家重要的白兰地企业，在收购了F.C.科利森（1833年成立）之后，它宣称自己是南非未曾中断生产的白兰地厂商中历史最悠久的一家。范瑞斯采用干邑的方式蒸馏白兰地，甚至还有自己的现场制桶工厂。

南非最大的白兰地生产商是凯樽汇酒业集团[①]，简称

与时俱进的圣格诗白兰地采用升级版酒瓶和酒标，在世界市场上彰显自己的品质。

---

① 其原名 Koöperatieve Wijnbouwers Vereniging van Zuid-Afrika 直译为南非葡萄种植者合作协会。

如图中范瑞斯酒厂的这只铜质阿伦比蒸馏器所示，南非的白兰地一直以传统工艺生产。

KWV。公司成立于1918年，1923年成为葡萄酒酿酒合作社，1926年开始生产白兰地。在历史上的不同时期，这家合作社有时是私人企业，有时是公营公司。1977年之前，作为南非葡萄酒行业的监管机构，KWV不得参与国内市场的竞争，因此它生产的瓶装白兰地一律用于出口。不过，KWV仍会将散装白兰地批发给其他公司，后者再将其陈酿装瓶。目前它是一家私营企业，产品兼顾出口与内销，在南非白兰地和葡萄酒市场上仍然占有非常重要的地位。

还有一家重要的公司叫克利普瑞夫特。1938年刚刚创立时，它只是一家小小的后院酿酒作坊，但其产品却以创纪录的速度崛起，成为南非最著名的白兰地。然而遗憾的是，后来南非白兰地的品质和形象一落千丈，而克利普瑞夫特白兰地恰恰是罪魁之一。到20世纪中叶，国产白兰地在南非的地位已恰如廉价朗姆酒在美国的地位。"克利普瑞夫特加可乐"就好似美国的朗姆酒加可乐一样，代表着一种最普通的低端饮料。随后又出现了"1—2—3"玩法（有时也叫"3—2—1"），即低配版的欢乐之夜：1升白兰地，2升可乐，外加一辆3升排量的福特车。

但在过去几年里，人们发觉这个行业的面貌开始发生改变。正如美国的朗姆酒品牌为了摆脱"朗姆酒加可乐"的形象而竭力提升自己一样，南非人也开始证明他们的产品值得在烈酒界享有更高的地位，并已经面向国内外市场推出了不少优质陈年白兰地。

成立于1984年的南非白兰地基金会已将白兰地的生

产规范为四个等级。第一级称为调配白兰地,生产目的是用于制作混合饮品,其中经壶式蒸馏器蒸馏并在橡木桶中陈酿三年以上的白兰地至少要占 30%,其余部分可以是未经陈酿的中性烈酒。第二级是年份白兰地,要求含有至少 30% 的壶式蒸馏白兰地,至多 60% 塔式蒸馏并熟成八年及以上的烈酒,以及最多 10% 的烈性葡萄酒(未熟成)。第三级是壶式蒸馏白兰地,必须含有至少 90% 的壶式蒸馏白兰地和最多 10% 的未陈酿中性烈酒。第四级是酒庄白兰地,从生产、陈酿到装瓶的整套流程必须在同一个酒庄内完成,酒标上除了注明白兰地的种类之外,还会印有"酒庄"的字样。

酒标上可能会使用 VS 和 VSOP 等术语,但它们的陈酿规定与干邑地区并不相同。在南非,白兰地必须在容量 340 升的法国橡木桶中至少陈酿 3 年。索雷拉陈酿法也是允许使用的。虽然南非没有指定的白兰地产区,但用于酿制白兰地的葡萄往往出自几个主要的酿酒葡萄种植区,包括伍斯特、象河、奥兰治河、布里德河和小卡鲁。

白兰地仍然是南非最畅销的烈酒。2008 年,全新年度盛典"精品白兰地荟萃节"正式开幕,其举办目的主要是吸引年轻消费者的关注,并为白兰地打造魅力十足的高端定位。有趣的是,如今南非白兰地基金会网站在追忆蒸馏烈酒的起源时,也用上了如此浪漫抒情的辞藻:"白兰地的生产就如同炼金术,能将土、气、水、火这些大自然赐予的元素点化成黄金。"

美国出产的白兰地

历史上，美国人曾将白兰地视作家庭必需品，因为它功能多样，比如既能作饮品，又可供药用；其实在20世纪之前，白兰地一直被划入药品之列。在那时，美国东部会进口干邑和其他法国白兰地，但在西部，白兰地却是甚为少见，市面上能买到的只有秘鲁皮斯科。

19世纪后期，西部人口出现爆炸式增长，那里随即燃起一股对白兰地的渴望。美国的两家老牌白兰地企业都是在19世纪80年代启动生产的：基督教兄弟创立于1882年，科贝尔创立于1889年。弗朗西斯·科贝尔是波希米亚移民，淘金热兴起后，他和他的兄弟们被加州的无限机遇所吸引，来到这里闯天下。他在旧金山北部找到一块土地，开始生产起泡酒。熟谙葡萄酒生产和营销的诀窍后，科贝尔又继续用同一套方法来打造白兰地。

在科贝尔公司创立数年前，世俗宗教组织基督教兄弟为了赚钱办教育开始酿制和销售白兰地，科贝尔可能就是受它的成功启发（也可能并不是）。基督教兄弟（以及进入加州商业白兰地行业的许多后来者）将自己的生产和蒸馏总部设在加州肥沃的内陆山谷中，这里也是该州大部分果蔬的种植地。在这片地区，刚刚起步的白兰地生产商可以买到大量价格低廉的葡萄，其中许多厂商在生产初期选用的只是普通的鲜食葡萄，例如汤普森无核葡萄和粉红葡萄。后来，该地区种植的酿酒葡萄多了起来，其中一些也用于生产白兰地。

白兰地生产商通常都会仿效干邑的原料筛选原则：他们

认为，为了生产出美妙绝伦的白兰地，就必须选用那些不可能酿出优等葡萄酒的葡萄。从某种程度上来说，这话没错。如果葡萄中所含的花香等芳香成分经过蒸馏后仍会保留下来，那么这个论断就尤为准确。而且这些品种的葡萄可以趁酸度较高、糖分较少时提早采摘，因此价格也更便宜——葡萄在葡萄藤上的生长时间越短，意味着每一季收获前用于打理葡萄园的时间和成本就越少。

　　加州商业白兰地的生产刚刚肇兴几十年，酿酒活动就因禁酒令而被迫中止，而干邑由于其所谓的药用价值，就成了禁令实行期间唯一被允许进口到美国的酒。其实从禁酒令颁布之前的数年开始，白兰地在美国就已不再被归类为药品，但许多医生仍然将其写进处方。就像今天的邦迪创可贴一样，白兰地当年也被视为家庭药箱中的常备物品。关于禁酒令期间美国的白兰地生产情况，我们并没有全面翔实的记录，但那个时候，家庭自酿白兰地的设备还是可以弄到的，只不过用来发酵的水果却只能有什么就用什么。

　　禁酒令几乎扼杀了加州的消费类葡萄酒行业，而在禁令废除后，元气大伤的白兰地生产也亟待复苏。在加州各地，有一些人瞅准这个断层期，重启商业白兰地的蒸馏和分销活动。这些生产商包括库卡蒙加山谷的乔瓦尼·瓦伊（1933年）、德拉诺市的安东尼奥·佩雷利-米内蒂（1936年）、弗雷斯诺市的乔治·扎尼诺维奇（1937年）以及欧内斯特·嘉露和朱利欧·嘉露兄弟（1939年）。嘉露公司之所以会进军白兰地行业，好像是因为在禁令解除之后的几年里，酿酒葡萄多次喜获丰收。一般来说，如果生产出的葡萄酒多得卖不完，余下的存货就会被蒸馏成烈酒，派上各种用场。未经陈酿的烈酒白兰地也被用来强化甜葡萄酒，这种酒在 20 世纪大部分时间里深受美国人喜爱。

起初，欧内斯特·嘉露为了帮衬一位身陷窘境的朋友，便从他那里购进了几千桶白兰地。随后，嘉露公司在 1949 年决定用自己"多余"的葡萄酒生产嘉露白兰地。1967 年，公司打造了一款名为埃登罗克白兰地的新产品，1973 年又在弗雷斯诺开办酿酒厂，重新推出嘉露白兰地。1975 年，外观高贵的 E&J 白兰地问世，埃登罗克遂告停产。公司选用的是干邑地区传统的鸽笼白葡萄，外加一部分白诗南、歌海娜、巴贝拉和麝香葡萄。1977 年，E&J 白兰地首度行销全国；此时，朱利欧·嘉露也开始生产自己的单品种葡萄白兰地。

禁酒令撤销后，基督教兄弟和科贝尔重新开始生产白兰地。为满足美国市场日益增长的需求，另一些大公司也纷纷涌入这一行业，其中有许多同时也是葡萄酒生产商，生产的还是 20 世纪中叶美国最热销的品牌，如爱玛登、意大利瑞士殖民地、保罗梅森等等。20 世纪 50 年代中期，加州有十余家厂商开始生产以葡萄酒为原料的时尚款系列白兰地。美国大型分销商也杀入这个利润丰厚的行业，有的为生产商投资，有的与之合伙经营，或二者兼而行之。其中规模最大的四家分销商是施格兰、辛雷、国家酿酒公司和海勒姆沃克父子公司。

为了将自己的酒同其他质量较次的白兰地区别开来，20世纪初期，加州葡萄酒行业的先行者欧内斯特·嘉露和朱利欧·嘉露兄弟为自家白兰地设计了全新酒标，上面仅印有"E&J"两个字母。

　　接下来的几十年里，加州白兰地的需求一直极为旺盛，于是那些规模最大的公司发现，将白兰地蒸馏后运到肯塔基州不失为一种变通之法，因为当地有一些非常宽敞的棚屋，里面满是可以用来陈酿白兰地的旧波本桶。虽然各家生产商都以陈酿作为宣传卖点，但对于这些"加州白兰地"究竟是在何处陈酿，它们却往往只字不提。

　　然而在产量增加的同时，产品质量却下降了。虽然许多公司最初是以手工方式生产白兰地，采用干邑风格的壶式蒸馏器完成部分或全部蒸馏工作，但后来大部分公司都改用大容量工业塔式蒸馏器（与雅文邑的小型手工塔式蒸馏器不可同日而语），因为它们必须紧跟市场需求，而此时的市场对质量并不是很敏感。20世纪中叶以后，加州白兰地的声誉毁失殆尽，这是多种因素共同作用的结果：国内产品质量低劣；年轻一代嫌弃父辈钟爱的酒饮；各个年龄段出国旅行的人都越来越多，而他们又都在国外品尝过质量上乘的白兰地。

但当时（以及现在）仍有一批美国消费者青睐风格粗犷的美国白兰地品牌，比如基督教兄弟、嘉露、科贝尔和保罗梅森。最终，美国白兰地生产商"四巨头"中的两家，连同它们在肯塔基州的合作机构，均被以生产经销波旁威士忌起家的企业集团收购：基督教兄弟被（名字恰如其分的）天堂山收购，保罗梅森被星座集团收购。然而，咎由自取也好，无辜受累也罢，当其中许多白兰地以低端价格出售后，这一行业的地位也每况愈下。到 20 世纪 80 年代，加州白兰地已经失去了年轻人和精英阶层的欢心。

这种局面持续了数十年之久，直到最近干邑人气大涨

加州葡萄酒公司科贝尔早年间生产的白兰地坚持采用传统风格的外包装，看起来跟药瓶差不多——别忘了，在长达几个世纪的时间里，人们一直认为白兰地可药用。

DISTILLERY OF **F. KORBEL & BROS.,** SONOMA CO., CAL
723 Bryant St., San Francisco, Cal. ᐧᐧᐧᐧ 40 La Salle St., Chicago, Ill.

This photograph shows an early Korbel calling card. Taken near the turn of the century, we can see the winery employees posing for this group shot in front of the old Brandy Tower. It's interesting to note that the Brandy Tower does not have the steel reinforcing rings which were added after the San Francisco earthquake and fire of 1906.

照片中是科贝尔在北加州的蒸馏酿酒厂，可以看出，早在1906年之前它就已经在旧金山建成很久了。

才出现转机。一些历史悠久的美国老牌白兰地趁势而起，纷纷开始改换包装，重新定位自己的品牌。例如，2003年嘉露开始用阿伦比蒸馏器生产白兰地；目前公司已占有43%的市场份额，而且由于近期年轻消费者和喜好白兰地的女性人数都有所增加，它正好借此东风大展宏图。

人们对美国白兰地的印象有所改观，可能是出于以下几种原因：一是各家企业推出一系列高端新品；二是工匠精神运动的蓬勃发展令白兰地进入公众视野；三是21世纪的调酒热潮使干邑和白兰地重新获得赏识。下文将对比做出进一步讨论。

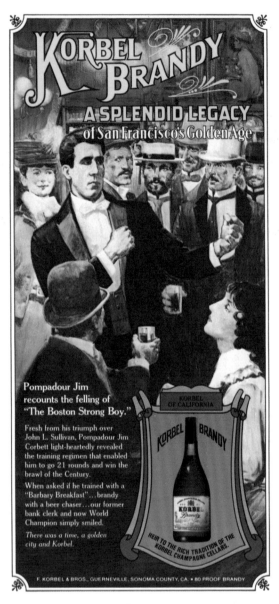

科贝尔白兰地以上流社会高档酒的形象现身；在这则广告中，
一位当红拳击手被品位高雅的绅士淑女们热情追捧。

［第九章］

干邑知识大百科

在前文中，让我们为之着迷的干邑历史故事刚刚讲到20世纪初，当时这一地区已经开始从摧毁其大量葡萄园的欧洲葡萄藤虫害中恢复元气。同时，干邑又遭遇了另一大威胁：19世纪末期，其他国家冒出了一批把自己的烈酒也称为"干邑"的白兰地生产商。

为了彰显自家白兰地独一无二的特质，干邑人首先尝试为干邑葡萄园圈定一片专属区域。然后，他们又对干邑的生产和陈酿做出了明文规定。再后来，他们力图与其他白兰地产区签订贸易协议，以期限制"干邑"一词的使用——这场充满艰辛的斗争至今仍在继续。

参阅该地区最全面的土地分析资料（从1860年开始）之后，干邑人于1909年划定第一版专属种植区。1936年，干邑成为AOC（原产地命名控制）地区，六大种植区（葡萄园区）也于1938年最终确定，即大香槟区、小香槟区、边缘林区、优质林区、良质林区和产地林区（亦称为普通林区）。"优质香槟"是一个附加的称号，但并不是种植区——它是指由大小香槟区的葡萄混酿而成的干邑，其中大香槟区的葡萄至少要占到50%。

这里的"香槟"意思就是"乡村"，为白垩质土壤，其葡萄园里种植的葡萄能酿出最上等的干邑。大香槟区葡萄园是公认的最佳种植区，小香槟区（分布着另一种白垩土）紧随其后，接着按优劣顺序排列，依次是边缘林区（黏土和砂质土居多）、优质林区和良质林区（土壤中白垩的含量和种类不一），最后是普通林区（砂质土比例较高）。

过去，定价最高的干邑必须用排名前二或前三的种植区所产葡萄酿制，但如今生产商也在尝试用其他产区的上等陈酿干邑进行调配，并且取得了成功。最近，卡慕酒庄推出了融合海盐、泥煤和威士忌式烟熏风味的雷岛系列干邑。尤为相宜的是，这个系列或许会让更多人了解干邑，因为在早期同荷兰人的贸易中，从雷岛采集的优质食盐正是促成干邑这种烈酒问世的因素之一。

在所有葡萄中，唯独白葡萄品种可以用来酿制干邑。干邑地区选用的葡萄主要有鸽笼白、白福尔和白玉霓。此外，干邑生产商可能还会用到最多 10% 的福丽酿、白朱朗松、梅利耶-圣弗朗索瓦、蒙蒂勒、塞莱特和赛美蓉（每个品种的比例都不能超过 10%）。如今，干邑生产商已经开始居安思危，考虑未来几年气候一旦发生剧变，哪些葡萄品种会表现最优。

干邑必须采用二次蒸馏法生产，所用陈酿桶必须是当地制作的上等橡木桶，其木料取材于附近的森林。当地的利穆赞橡木质地松散，人们认为能够为白兰地析取出最理想成分的木材非它莫属。橡树从森林中被精心挑选出来，之后木材被切割成段、劈分成桶板，并自然风干两到三年。

这些昂贵的木桶都是手工制作而成。对于木桶内壁的烘烤程度，每家干邑生产商可能都有自己的选择：有的喜欢浅色，有的倾向于深色，这取决于它们希望为自己的烈酒赋予怎样的感官特质。（虽然干邑的大部分颜色最好是在桶陈过程中形成，但法律也允许在最终调配时添加一些焦糖色，以及白糖和一种名叫博伊兹①的木本"茶"。）在自然陈酿的过程中，干邑会从晶莹剔透的无色液体依次渐变为黄色、金琥珀色和黄褐色，当陈酿期达到十年时，酒体就会呈现出浓郁的桃木棕色。

在整个过程中，生产商还必须把蒸发的因素考虑进去。酒液每年会蒸发掉 2% 至 6%，这一部分被称为"天使之享"，用法语说就是 la part des anges。干邑市的年度慈善拍卖会也以此为名，每年都会吸引国际高端买家前来竞拍。

干邑会在干燥酒窖和潮湿酒窖两种环境中陈酿。为了使所需的香气和味道达到最佳，有时酒厂会制定一套复杂的方案，每隔一年左右就将干邑挪动一次（转移的通常只

这张照片展现了 25 年陈酿过程中干邑颜色的变化：从无色透明逐渐过渡到稻草金、琥珀色和棕色。

---

① 原文为 boisé，干邑的一种添加剂，是用木片（例如橡木片）煮水后浓缩而制成的深褐色黏稠液体，目的是增加干邑的木香味和陈酿感。

是酒液而不是木桶）。在潮湿的酒窖里，酒精的蒸发速度更快一些，干邑会变得圆润柔和，保留更多的果香和花香。而在干燥的酒窖中，水分蒸发得相对较快，在这种条件下，干邑中会含有更多的辛香和木香味。

酿酒师和酒窖主管都有自己的独门秘籍，为自家干邑打造出与众不同的口味名片。这些能带给人感官享受的元素包括花香和柑橘香，烘焙香料和雪茄盒的香味，另外还有太妃糖、咖啡、雪松、皮革、干果和香草等无数其他香气成分。

虽然干邑生产方法实现了标准化，但并不是说从此以后就一帆风顺了。二战期间，干邑的大部分地区都被占领，不过富有远见的干邑生产商仍设法保住了大部分库存以待将来。可他们究竟是如何做到的？相关故事通常并不会透露给来访的游客，因为据说一些生产商可能同魔鬼做了交易。如果有人好奇个中细节，不妨到讲述干邑历史的长篇巨著中去一探究竟。

二战后，干邑局（即法国干邑行业管理局，简称BNIC）于1946年成立，负责监督干邑在国内和全球的生产销售，其全体委员会由17个葡萄种植园和17个干邑酒庄组成。根据BNIC的规定，用于生产干邑的葡萄必须先制成葡萄酒，然后不经陈酿，尽快蒸馏成"生命之水"。在干邑地区，蒸馏须得在葡萄采收次年的4月1日之前完成。4月1日蒸馏截止期限后，干邑还要经过至少两年的桶陈，否则不得向公众出售。

著名干邑酒庄御鹿有一项自己引以为傲的传统，即坚持生产一种20世纪中叶之前极为常见的干邑。这种风格的干邑名为"先抵达"，它被装进酒桶中运到英国，然后在当地仓库中陈酿后装瓶。英格兰全年气候凉爽潮湿，酒液蒸发量较少，因此"先抵达"干邑的风味与在干邑地区陈酿的烈酒有所不同。（过去50年里，御鹿一直是女王伊丽莎白二世钦点的干邑供应商。）

需要注意的是，干邑的一切变化都是在桶陈过程中发生的，一旦装瓶即可直接饮用。装瓶后的干邑品质不会再继续提升。即便在酒桶中，干邑的变化也会在某一刻停止，这一刻可能是在几十年后，最远可达80年。到了这个时候，如果干邑还没有装瓶待售，酒窖主管就会将其转移到一只叫作"珍妮女士"（又名"黛米约翰"①）的大号球形玻璃容器中，再把它珍而重之地锁进一间屋子，这里面保存的都是生产商最宝贵的干邑。酒窖的这片区域有个很贴切的名字——天堂。

干邑开瓶后，可以在阴凉避光处保存数月乃至一年。干邑的最佳赏味期因酒龄而异，酒龄较短的干邑可能适合在装瓶后数年内饮用，而对于陈年佳酿，这个时间则可以延长至数十年。不过到了最后，哪怕是瓶装干邑，其味道和香气也会开始逸散。

当前的干邑等级要求是于1983年最终确定的，虽然未

---

① 用柳条包裹的细颈大肚玻璃瓶，其英文为demijohn，可能就是来自法语dame-jeanne，即珍妮女士。

来修订版正在讨论当中，但并不一定会落实。目前，干邑局网站上列出的等级有：VS（非常特别）或三星级为计数2，即至少经过两年桶陈；VSOP（特优浅色陈年酒）或珍藏级意味着计数4，即桶陈至少四年；拿破仑、XO（特级陈酿）或Hors d'âge代表计数6，桶陈至少六年。干邑等级反映的是调配酒液中最年轻白兰地的酒龄。而最近开始流行起来的年份干邑，其所含烈酒必须全部由指定同一年份采收的葡萄酿造而成。市面上出售的大部分干邑（约占85%）都是VS或VSOP，其余则是酒龄更长的干邑。

　　"拿破仑"一词在概念上存在一些混淆，因为在过去，只要干邑（以及世界其他地区）的生产商想表明自家白兰地是陈年老酒，就一律称之为拿破仑。如今干邑地区正式规定，"拿破仑"一词是指干邑的陈酿时间至少为六年，同XO干邑一样。更复杂的是，在实际应用中，酒标上注明"Extra"字样的干邑通常比XO级陈酿时间更长。许多XO和Extra干邑中都兑入了大量酒龄老得多的干邑，且每家干邑酒庄调配出来的酒液，其所含确切成分都是不同的。最近，为了吸引各种类型的消费者，一些干邑酒庄还推出了拥有专属名称和独特酒标的特别版干邑。

　　许多国家把VS、VSOP和XO等指定称号挪为己用，甚至连不再使用"干邑"一词的国家也不例外。这种标称法很方便，因为如果产品的确符合该命名体系的标准，那么买家就可以据此判断酒龄的长短。然而，每个白兰地产区都各有一套陈酿和标称规则，因此，产自其他地区的

VSOP 未必与干邑的 VSOP 酒龄相同（生产过程可能也不会那么考究）。

目前列入干邑管理局名录的干邑生产商共有 325 家，小公司和大企业、个体户和合作社都包括在内。但其中有四家大企业的名字在全球家喻户晓，令其他大多数厂商黯然失色，它们就是拿破仑、轩尼诗、马爹利和人头马，其产量之和占干邑总产量的 85%。这些企业旗下有一些葡萄园，它们会使用种植者供应的葡萄酒蒸馏出一部分干邑，同时也会依照惯例向种植者购买现成的蒸馏酒，然后进行陈酿和调配，装瓶后推向市场。

每家干邑生产商可能都会留存年份跨越数十载的陈年干邑样品，一是用作参考，二是可以用来为一些特殊场合调配理想佳酿。

如今的酒瓶大多是圆柱形或烧瓶形，但也有许多玻璃酒瓶拥有各种曲线玲珑的外观轮廓。生产商为自家的顶级干邑定制各种美不胜收的雕塑艺术造型，而达到这种层次的酒瓶，本身也令干邑身价倍增。酒标的式样也千差万别、各具特色，有的散发出古老而传统的韵味，也有的表现为纯粹的后现代风格——酒瓶正面最多不过两个字。总而言之，设计可以极大地提升干邑的声望和价格，一些顶级瓶装酒竟然可以卖到数千乃至数万美元。例如，过去 5 年间，单瓶轩尼诗、御鹿和拿破仑都曾卖出过 10000 美元（6000英镑）的天价。

当人们想要探索一种优质烈酒的奥秘时，很容易从最便宜或酒龄最短的酒开始尝试。不过对于干邑来说，这样做却并不妥当。VSOP 才是最适合入门的干邑级别，因为你不必破费太多，就可以体验到这种烈酒的细腻香气和顺滑口感。哪怕是打算用干邑来调制鸡尾酒，调酒师也会告诉你，从 VSOP 着手仍是最佳选择，因为它能与调酒饮料和其他调味料融合得异常完美。

VSOP 干邑的起售价约为 30 英镑（45 美元），各酒款的具体价格以此为基础适当上调。当然，相比之下 VS 干邑要便宜一些。顶级酒庄出产的名牌 XO 大部分都是从 60 英镑（100 美元）左右起售，而知名度较低（有时声誉也稍逊一筹）的干邑酒庄，其 XO 的定价可能会低得多。

通常来说，消费者打算购买干邑时，价格是需要考量的重要因素。不过市面上除了 VS、VSOP 和 XO 级别的干

早在亚洲最近一个龙年尚未到来之前，金露干邑就已打造了一条名为"龙"的产品线，所有产品均为陈酿期至少40年的干邑。这个系列的酒每瓶都用木箱独立包装，对亚洲消费者尤其具有吸引力。

邑，还有各种以专属名称生产的干邑，其中有一些上架时间很短，靠稀缺性来提升自己的吸引力。这些限量版干邑大多价格昂贵，除了设计别致的酒标酒瓶之外，还会凭借与众不同的香气和味道带给人别样的感受。可能这一家公司突出的是辛香风味，而那一家的主打卖点则是融入烈酒的浓郁果香，这些都取决于它们想要吸引哪一类顾客。

说到饮用干邑，很多人都看过一幅老漫画：一位富绅手举一只硕大的球形餐后酒杯，炫耀着杯底那浅浅一层黄褐色液体。其实，这种大肚球形玻璃杯早就过时了。50或100年前它的确用处不小，因为当时的室内要比现在冷得多，而存放白兰地的酒窖温度通常更低，于是，为了释放干邑的微妙香气和味道，人们就习惯用手掌的温度来加热杯中液体。加热时，香气会逐渐飘溢出来，并在球形杯的杯口处驻留一段时间，只需轻轻吸嗅即可品味其芳美。

但如今，我们不必再这样做了。在现代房屋中，室温下储存的干邑一开始就达到了可供饮用的温度。将干邑斟出后，即刻凑到杯边吸气，芳香便会扑鼻而来。此外，用晃动酒液的方式来释放香气也不甚可取，因为只需轻轻旋转玻璃杯，清雅细腻的香气就会从微温的干邑中飘散出来。

正式的干邑品酒杯尺寸小巧，中部较宽，杯口处稍稍内收，将干邑的香气拢住，让人充分品味到这种烈酒的芳香。

在干邑地区，如今喝干邑都流行用小酒杯，尺寸同小号白葡萄酒杯或雪利酒杯相仿。（这可以说是一种复古时尚：早年间的干邑酒杯体形很小巧，大概就是仅为饮酒而造，而不是供人慢条斯理地一边温酒一边鉴赏。）用来啜饮的陈年干邑宜少量斟出：每次 20 至 40 毫升，也就是 1 液量盎司左右。

在选购干邑时，我们不难发现，占干邑地区产量 85% 的四大干邑酒庄也夺取了全球销量的大半江山。它们中的大多数都有大型企业为后盾，这些企业以全资或控股的方式拥有这几家酒庄，为其提供全球扩张、产品开发、分销和营销所需的资源。轩尼诗归于路威酩轩（酩悦·轩尼诗-路易·威登集团）旗下，马爹利是保乐力加的品牌，拿破仑为比姆全球（现已被三得利收购）所有，而人头马属于人头马君度集团。根据 2012 年 3 月的报道，这四家公司上一年度的全球销售总额约为 30 亿英镑（50 亿美元）。

除了上文提及的品牌之外，值得品尝的优质干邑还有许多，我们不妨根据市场供应情况和饮酒场合（当然还有个人财力）选购一番。这些品牌往往也是比较容易买到的：ABK6、伯爵斯云利、豪达男爵、百事吉、卡慕、魔术、蒂乐仕、德拉曼、费朗、凤马、法拉宾、古殿、哈迪、御鹿、菲利傲、约翰逊、朗帝、路易老爷、墨高、诺曼丁-梅西耶、贝雷特、普诺尼。

干邑鸡尾酒与21世纪的趋势

在 21 世纪，干邑形象的核心就是奢华。无论是净饮还是调制成高端鸡尾酒，抑或与调酒饮料混饮，喝干邑的人追求的都是一种极致奢华的体验。在英美以及紧跟西方文化潮流的国家，的确有一部分人仍喜欢在晚餐后净饮高档白兰地。在这个层面上，干邑同奢华生活的关联仍然深入人心，而在其中占据主导地位的就是顶级干邑。不过，如今白兰地的复兴是多股潮流共同作用的结果，即东亚文化、美国流行音乐和华丽炫目的调酒艺术。

20 世纪末，有两个文化群体的干邑和白兰地消费量同时开始激增。其中一个在亚洲，尤以中国香港高端消费者为最；另一个是美国的城市（内城区）消费者。当时，在全球其他大部分地区，干邑的增长已基本陷入停滞，于是干邑生产商开始更关注这两个地区，结果自然是收获颇丰。

在美国城市中，白兰地和干邑的消费量在西语拉美裔和黑人聚居区中增长最为显著。就拉美社群来说，这可能是早已流行于墨西哥和其他拉美国家的西班牙白兰地文化在这里进一步传播的结果。而在城市黑人社群中，这一现象有多种解释。有人说是由于当代年轻人的叛逆和标榜个性，也有人认为同二战中黑人士兵在欧洲的经历有关——不过按照后一种说法，人们对白兰地兴趣大增的原因都可以追溯到 20 世纪更早的时期。实际上，早在都市说唱运动开始为白兰地发声之前几十年，内城区的白兰地消费量就已经相当可观了。而世界上其他许多地区却是直到 2001 年才开始了解白兰地，因为这一年布斯塔·莱姆斯与吹牛老

在干邑鸡尾酒峰会的鸡尾酒调酒大赛中，一位调酒师调制出一款火焰鸡尾酒。

爹联袂演唱的歌曲《传递拿破仑干邑》大火，随后二人又推出这首歌的混音版和音乐短片，同样大获成功。

其后十几年间，有150多首新发布的说唱歌曲提到了干邑品牌，其中在歌词里出现频率最高的就是拿破仑和轩尼诗（轩尼诗有 Hennessy 和 Hennessey 两种拼写方式）。与此同时，为了全力经营这块市场，一些干邑生产商还同几位知名艺人建立了合作关系。卢达·克里斯的魔术干邑就是同哈特曼干邑公司老板金·哈特曼联名推出的，不过在广告宣传活动中，卢达·克里斯仅以自己的真名克里斯·布里奇斯亮相。德瑞博士与 ABK6 干邑酒庄共同发布了自己的"余波"干邑，说唱歌手 T.I. 和马爹利干邑也宣布结成合作伙伴——但几个月后，T.I. 银铛入狱，马爹利随即与之解约①。

一些顶级的干邑公司还生产专供城市高端交易的干邑。音乐人杰斯将百加得出品的铎世干邑引入美国，豪达男爵干邑也是该公司旗下的产品。人头马最近推出了一款名为"城市之光"的限量版 VSOP 干邑，其酒标图案在紫外光照射下会散发出红色微光。每年问世的高端干邑都越来越多。一个有趣的现象是，干邑公司有时并不会将此类产品发布到自己的官网上，它们更希望消费者去访问以这些干邑的专属名称命名的网站。此时此刻，干邑生产商似乎在同时各出奇招，探索如何维系这种经典烈酒在全球的奢华吸引力。

让我们把场景切换到亚洲。在这里，干邑的消费只能用一个词来形容：爆炸。很多人都听说过，在如今的中国，顶级波尔多葡萄酒是高端消费者的标配，但并不是每个人都知道，眼下干邑在烈酒界已经获得了堪与波尔多媲美的地位。日本顾客最倾心的西方烈酒是苏格兰威士忌，但中国和亚洲其他地区的消费者却是白兰地的狂热拥趸。在中国，白兰地一词就等同于 XO、Extra 和年份酒：只有极品佳酿才有资格入选，而且酒越陈越好，存世量越少越好。

2012 年，中国干邑市场的价值已超过昔日最大的市场——美国。许多干邑生产商被这惊天增长率所吸引，在过去几年纷纷瞄准亚洲发力。而且，由于中国人往往只买

---

① 原文有误，与 T.I. 合作的实际上是人头马。

最高端的产品，这种形势在可预见的未来仍有望持续下去。目前有为数众多的干邑酒庄专为中国市场打造特制的酒瓶包装、调配酒和酒标——至于它们在世界其他地区是否亦作如此宣传，那就另当别论了。

最近，有媒体发文质疑干邑是否能继续满足中国市场的需求，因为精明的中国进口商和消费者又发现了雅文邑。这自然令雅文邑地区欣喜不已。由于地处内陆，雅文邑的名贵白兰地在长达数世纪的时间里一直屈居次位，但现代交通运输已经克服了地理障碍。如今的雅文邑在中国享有当之无愧的盛名。

但中国人现在又继续探索，将目光投向其他生产正宗（即以葡萄酒为原料的）白兰地的市场，距离最近的就是澳大利亚。对于新市场而言，从干邑以外的地区采购白兰地很简单，因为全球大部分生产商在白兰地酒标上用的都是同一套分级标准：三星或 VS 通常代表第一级，VSOP 是第二级，XO 是第三级。此外，澳大利亚也会使用拿破仑、特级陈酿和年份酒等术语，特别发售版则有自己的专属名称。虽然同一等级白兰地在世界各地的陈酿时间不尽相同，但这种标称方式却能让新经销商和消费者在初探白兰地世界时轻松入门。

其他亚洲国家也在追逐这一时尚。一些国家是受早期殖民者的影响开始养成饮用白兰地的习惯；而在另一些国家，人们则是直接效仿自己在音乐短片中见识到的都市生活——当今菲律宾年轻人喜爱白兰地正是受此影响。不过，该国上几代人还同时保留着饮用西班牙白兰地的传统，派卓多美生产的芬德多赫雷斯白兰地是他们的至爱。

除中国以外，其他亚洲国家也进口一些白兰地。例如，越南在20世纪90年代末是全球第五大白兰地消费国，如今消费势头依然强劲。喝白兰地的主要是20至50岁的男性，干邑是他们最偏爱的烈酒。在这些男性当中，选择其他白兰地的不到10%，其中约有1%钟情于雅文邑。顶级干邑酒庄的产品最受追捧，因为这种烈酒大多是年轻人在夜店等场所当众购买饮用的。

俄罗斯和其他亚洲国家也生产白兰地，不过，由于其中一些地区种植的葡萄数量有限，它们生产的许多白兰地其实并不是葡萄酒蒸馏的产物。马来西亚的白兰地产业很发达——事实上，全球最大的白兰地品牌就是马来西亚的皇胜。但在这里，未曾明言的那个因素才是关键：制作白兰地的原料。只有在部分情况下，皇胜才是由葡萄酿制而成，而选用的葡萄又只有在部分情况下才是酿酒葡萄（与之相对的是鲜食葡萄）。

不过，有些市场现在倾向于追求货真价实。如今，俄罗斯的法纳歌利雅酒庄以及这片大陆上其他多家白兰地生产商都在强调，它们的白兰地是经橡木桶陈酿的，法纳歌利雅最近还自己开办了一家现代化制桶厂；干邑地区的布鲁瓦酒庄是俄罗斯肯氏集团旗下产业，集团从那里进口蒸馏烈酒和陈年干邑，同时也在国内生产调配白兰地；印度梦神白兰地选用的葡萄部分在印度本土种植，部分从法国进口；

而一位在中国工作的顾问说，中国正在尝试采用传统古法蒸馏白兰地。

在很多地方，"白兰地"不过是贴在有色蒸馏酒包装瓶上的一种标签而已。只要能行得通，随便什么农产品都可以拿来酿制白兰地，有时甚至会用菠萝充数。这种做法的优势是生产商降低了成本，顾客也可以省下不少钱：这些白兰地的售价还不到进口白兰地和干邑的一半。

关于白兰地产品，还有个有趣的趋势值得一提："白色"白兰地。从雅文邑到加利福尼亚，这种趋势好像随处可见。在烈酒行业，"白"实际上就是指澄澈透明的烈酒，但若用这个词来形容白兰地，似乎又有些自相矛盾之处，因为白兰地就是以赋予其美妙香气、味道和颜色的木桶陈酿过程而闻名。白色白兰地之所以看起来晶莹剔透，要么是因为它没有经过陈酿，要么就是橡木的颜色在陈酿后被滤出。

白色白兰地也是雅文邑的本土产品——实际上，这里的传统"白雅文邑"已于 2005 年获得正式 AOC（原产地命名控制）资质。酿制白雅文邑必须使用产自指定葡萄园的葡萄，品种包括白福尔、白玉霓、巴科和鸽笼白。蒸馏必须及早开始，在这之后，酒液要经过 3 个月的沉淀期或"熟成期"。接下来，生产商就可以向烈酒中兑水，将过高的酒精度降低，使其符合装瓶标准，即 40% 左右。在实际操作中，这一步骤通常要用 3 个月以上的时间逐渐完成。

白色白兰地抓住了一个大好机会：过去几十年里，用无色烈酒作鸡尾酒基酒成了流行趋势，这股时尚之风愈刮愈猛，而且看起来丝毫也没有减弱的迹象。白色白兰地的潮流甚至席卷到了干邑界：2010年，人头马用干邑生产出一款不经陈酿的无色烈酒"V"；生产黑金刚干邑①的轩尼诗也推出了一款"纯白"干邑，在多个国家试卖。严格来说，这些产品都不能被称为"干邑"，因为干邑必须经过一定时间的桶陈。然而，说不定市面上很快就会出现更多的白色白兰地，部分原因是，如果厂商想销售干邑地区的白兰地，却又不愿意遵照传统陈酿法的要求为之一等多年，那么直接卖白色白兰地就不失为一个好办法。

其他地区也有白色白兰地面世：南非的科利森公司生产了一款白金白兰地；美国的基督教兄弟则带着它的"冰霜"白兰地加入竞争。"冰霜"是经橡木桶陈酿的，但随后要进行一定的处理加工（并调味），所以瓶中酒液看起来清澈透明。以上这些无色烈酒均建议冰镇后饮用——为了紧跟时尚调酒潮流，也许可以用它代替伏特加来调制鸡尾酒。而正是在调酒这片大舞台上，干邑才真正开始火了起来。

---

① 轩尼诗黑金刚是为庆祝美国总统奥巴马任职的特别版干邑，最开始只在美国上市。它由35到45种不同风味的酒水混合而成，在法国橡木桶中陈酿至少5年。

小批量生产的白兰地和干邑

早在 21 世纪初鸡尾酒文化令干邑重焕生机之前，白兰地复兴之象就已在全球各地初现端倪，尤其是在 20 世纪 80 年代的美国，在工匠精神运动的激励下，一批新的白兰地生产商开始登上竞争舞台。然而后来，两家最著名的新厂商中有一家成为美国手工白兰地的业界翘楚，另一家却已黯然离场。

1982 年，干邑地区的知名干邑公司人头马同杰克·戴维斯建立合作关系，后者来自美国加州纳帕谷的起泡酒厂世酿伯格酒庄。他们在横跨纳帕和索诺玛两地的卡内罗斯葡萄酒产区开办了一家酿酒厂，为之取名 RMS 并寄予厚望。然而，酒厂以正宗干邑酿酒法生产出的白兰地在美国消费者当中根本没引来多少关注，更别提收获盛赞了——即使后来它更名为卡内罗斯阿伦比，情况也未见改观。

同样是在 20 世纪 80 年代初，美国人安斯利·科尔和干邑人于贝尔·热尔曼-罗班在森林繁茂的加州北部偏远小县门多西诺成立了热尔曼-罗班酿酒厂。虽然它生产的白兰地一向好评如潮，但美国人对此仍然兴致寥寥。这两家公司都坚持开办了 15 年以上，然而到 20 世纪 90 年代末，RMS（即后来的卡内罗斯阿伦比）已经关门大吉，热尔曼-罗班却终于实现了赢利，直至今日仍在生产手工酿制的烈酒。

科尔和热尔曼-罗班开创了一种为自家白兰地筛选葡萄的新方法。起初，他们采购的葡萄都是干邑地区所用的葡萄品种。但二人发现，在加州门多西诺县酿出的白兰地，口味与他们原先追求的并不一致，因为种植葡萄的土壤不

20世纪末，一位加州人和一位法国人整合双方资源，在北加州建立了热尔曼-罗班白兰地生产公司。首批产品的酒瓶设计低调内敛却又卓尔不凡。

同于干邑的那种白垩质土。他们意识到，自己脚下这片土地正是绝佳的酿酒葡萄产区，于是便决定碰碰运气，尝试用本地最好的葡萄酿制白兰地。经过无数次实验，他们终于掌握了用这些葡萄酿出优质白兰地的诀窍。不过，科尔和热尔曼-罗班不得不说服与他们签约的农民将采摘时间稍稍提前，因为这样才能恰到好处地保留酿制白兰地所需的酸度。如今，热尔曼-罗班酒厂会选用一种干邑葡萄，即鸽笼白，但主要依赖的是黑皮诺。根据收获情况，他们的白兰地也会用到本地种植的赛美蓉、长相思、仙粉黛、白诗南和麝香等葡萄品种。

选定最佳原料之后，科尔发现他必须向公众普及一下自己的白兰地内含哪些成分。他注意到，美国人将白兰地视为一种静态商品：他们在购买时只认牌子，而且认为价格越高质量就越好，对于白兰地的确切成分却不甚了了。于是，科尔慢慢引导美国公众，帮助他们加深对白兰地的了解。

在门多西诺取得成功后，酿酒师于贝尔·热尔曼−罗班开始同亚洲和世界其他地区的白兰地生产商探讨合作事宜。在当前干邑和白兰地热潮的鼓舞下，许多公司都迫不及待地想建厂酿制白兰地。然而，其中一些公司并没有适合蒸馏的良种葡萄——有些甚至根本没有葡萄。因此，于贝尔常常会发现自己置身于一个充满挑战的新世界，探索之旅妙趣横生。

再把视线转回美国。还有几家生产商也决定下场一搏。加州圣克鲁斯地区的丹尼尔·法伯创办了奥索佳酿酒公司，他自称加州白兰地的"第1.5代"。成立奥索佳之前，他曾前往法国和西班牙学习白兰地的生产和陈酿技术。后来，当他在20世纪80年代开启酿酒生涯时，还不知道RMS和热尔曼−罗班酒厂也刚刚成立。他只是一门心思地想做出世界一流的白兰地。如今，他尊于贝尔·热尔曼−罗班为业界先驱，称赞他以远见卓识创造出"加州阿伦比式"白兰地。

法伯采用干邑式蒸馏法，选取加州本地的黑皮诺、赛美蓉和鸽笼白等葡萄品种，每年的配方组合都稍有不同。法伯认为，奥索佳白兰地的风格介于雅文邑和干邑之间。他曾尝试用美国橡木桶来陈酿自己的白兰地，但随后意识到，法国利穆赞橡木的品质要好上许多。虽然他会在每一款年份酒陈酿结束后立即将其推向市场（一般不会进行调

加州热尔曼–罗班白兰地公司的首席酿酒师有着迷人的忧郁气质和法式复姓[1]，这与公司的形象堪称天作之合。

---

[1] 复姓（形容词）的英文是 double-barrelled，其中 barrel 一词也有"桶"的意思。

配），但有一种理念却令他心醉神迷：白兰地和人类的生命跨度其实是大体相当的，都可以长达80年。

杰普森是诞生于门多西诺的另一种加州白兰地。1985年，鲍勃·杰普森在俄罗斯河附近购置了一处地产兴建蒸馏酿酒厂。2009年，酒厂被现任老板们收购，改名为贾克森–基斯酒庄，并继续沿用杰普森的品牌名称在庄园中蒸馏白兰地，葡萄选用的是酒庄名下葡萄园中种植的鸽笼白。

我们再回过头来，看看卡内罗斯产区的情况。2002年练习曲酒庄接管了之前卡内罗斯阿伦比酒厂（原名RMS）的产业，同时也收购了尚在木桶中陈酿的白兰地。目前，公司销售的一款练习曲XO白兰地价格昂贵，选用当地的黑皮诺、鸽笼白、白诗南、帕洛米诺、霞多丽、白玉霓、麝香和白福尔葡萄酿制，陈酿时间为20年，由公司的葡萄酒酿酒师调配。

从小型手工制造企业发展起来的查尔湾是另一家位处卡内罗斯产区的酿酒厂，由卡拉卡舍维奇家族创建。让酒厂引以为傲的是，现任酿酒师已是其家族酿酒技艺的第13代传人，他父亲是1962年从巴尔干半岛移民到美国的。1983年，这家家族企业成立于斯普林山，葡萄酒产区纳帕和索诺马即以此山所在的山脉为界。公司在其生产的伏特加大获成功后开始扩张，几年前推出用白福尔葡萄酿制的83号白兰地——这是已经陈酿了27年的一款烈酒。

毫无疑问，就在读者阅读本章的同时，又会有许多由小酒厂创办的白兰地企业在全美各地涌现出来。位于纽约州北部乡村地区的手指湖酿酒厂就是其中新近成立的一家，厂主名叫布赖恩·麦肯齐。酒厂恰好坐落在手指湖葡萄酒产区的中心地带，因此，起初麦肯齐也生产以葡萄为主要原料的其他烈酒。但他对白兰地格外感兴趣，因为以手工方式生产白兰地的酒厂屈指可数。几年前，公司初创之际，他便开始蒸馏白兰地。2012 年，首批产品达到了他的陈酿标准，旋即面向市场发售，其酿酒原料取自当地出产的葡萄，包括琼瑶浆等本土品种和杂交品种。

还有一位著名的年轻酿酒商是常驻干邑地区的埃马纽埃尔·帕图劳德。他参与了一个有趣的小型项目，牵头人是为数不多的几个家族，他们长期种植葡萄并酿制葡萄酒供应给大型白兰地酒庄。但几年前，干邑地区的一小部分葡萄种植者开始承担额外的风险，自己陈酿、贴标并销售干邑。他们希望，随着公众对干邑和一切手工艺的兴趣越来越浓厚，家庭自制干邑也会更受欢迎。他们的这种转变同香槟区"种植者兼生产者"的创举相映成趣，而后者大约从 20 世纪与 21 世纪之交就开始受到外界的关注和好评，如今已取得巨大成功。

埃马纽埃尔同其父兄一起，在一处小型农场劳作，1934年，他的祖父就是在这里拥有了第一台简易蒸馏器。他们种植葡萄、酿造葡萄酒，蒸馏后将其中的一部分卖给人头马酒庄。人头马要求他们将葡萄酒带着酒泥一起蒸馏，为的是得到"更加圆润浓郁的风味"，而他们也以自己为这家大公司供应的产品为傲。不过现在，埃马纽埃尔也会专门留出一些葡萄酒自己蒸馏，打造自己的手工干邑。

　　帕图劳德家族用自家的老酒桶来陈酿和储存他们为自己调配的干邑。每年，他们都会购置四五只价格高昂的新酒桶。这些桶在购入之后的三年内都可以被视为"新桶"，也就是说可以在干邑陈酿期中关键的前 6 个月使用。帕图劳德家族将三四种干邑兑在一起，最终调配出融合各种干邑风格的酒液。过去，家族会留出一小部分干邑陈酿装瓶供自己饮用，而如今他们的瓶装酒开始对外销售，所以也会生产多种不同的风格。埃马纽埃尔重新接手家族生意之前，帕图劳德家的人从未生产过 XO，不过由于现在这种酒需求旺盛，他们也开始酿制起来。

在干邑地区，人们会使用一种名为"酒贼"的传统玻璃工具，定期对一桶桶陈酿中的干邑进行采样。一缕阳光恰到好处地照射在这份样本上，便于为帕拉齐的公司评定其口味、香气和外观。

最近，尼古拉·帕拉齐来到干邑，并带来了为其全球高品位顾客供应原料的优质合作伙伴。图中正在手工装瓶的是一款专供私人消费的定制版勾兑酒。

干邑地区大约有 1000 家家族酿酒企业，但由于家中的成年子女不愿继续留在乡间劳作，这一数字正逐年减少。虽然也有些年轻人想在干邑创立一家小企业，兼做种植与蒸馏，但遗憾的是他们往往财力有限，难以负担成本。埃马纽埃尔将这一现状归咎于地价上涨。他认为，在中国高端顾客和美国说唱音乐人的带动下，干邑的需求急剧攀升，结果在过去 5 年间，干邑全境葡萄园的价格均被大肆哄抬，高得几乎令人望而却步。

家族陈酿干邑进入市场的另一种方式就是经由公司代理，比如尼古拉·帕拉齐的 PM 烈酒公司。现居纽约的帕拉齐由祖父母在波尔多抚养长大。2008 年，他创办了一家定制化小批量生产的干邑企业。帕拉齐的第一笔生意从买下祖父一位朋友想出手的干邑开始，此后，他很快被介绍给其他一些境况类似的人：也许是某个家族想要退出干邑行业；也许是有些人恰好需要钱；也许只是有人想趁着干邑如今在全球爆红小赚一笔。甭管是什么原因，总之帕拉齐弄到了不少陈年干邑。他把这些酒收集起来，如有必要还会继续陈酿，然后装瓶出售给纽约和其他城市的精英顾客。

帕拉齐还多做了一步：有的人希望得到自己的独家专属干邑，于是他就为他们量身打造定制版的调配干邑和酒瓶。他同顾客一起（在自己家中或者在干邑），一次又一次细致周到、不厌其详地品酒，再设计酒标和别致的手工吹制酒瓶。最后，这些独一无二的调配干邑会被收藏家买下或供特殊场合使用。

有一首流行歌曲名叫《往日种种重焕新生》，这话用来描述今日的白兰地世界真是再合适不过。

鸡尾酒配方

下面是用干邑或白兰地做基酒的一些最著名的鸡尾酒配方，还有几个是新配方。经典配方中的配料比例可能并不符合如今鸡尾酒爱好者的饮用习惯，因此，我们按照现代人的口味对一些经典鸡尾酒略作改动，比如用赫雷斯白兰地代替干邑。鸡尾酒品鉴行家一定会借此机会好好享受一番，因为品尝过这些鸡尾酒之后，他们的经验领域会随之拓宽。

白兰地亚历山大

随便找一个人，请他说出一款白兰地鸡尾酒的名称，那么他脱口而出的应该就是这一款。不过现如今，真正了解它的人其实寥寥无几——当然，调酒师不在此列。20世纪30至70年代，这种酒在美国非常流行，但具体情况因地而异：在南方，白兰地是男性喜爱的饮品；而在北方，由于这种鸡尾酒泡沫丰富，喝它的人以女士居多。以下配方来自嘉露酒庄，它生产的白兰地或许可以用来调制美国20世纪中期的许多鸡尾酒。

30毫升（1液量盎司）白兰地

30毫升（1液量盎司）奶油

30毫升（1液量盎司）深可可乳酒

肉豆蔻粉

将上述液体配料加冰充分摇和，滤入鸡尾酒杯后撒上肉豆蔻粉。

白兰地可冷士

好笑的是，赫雷斯有一家新近成立的酒庄，名字却叫作"传统酒庄"。酒庄在整个赫雷斯地区四处收购酒龄较长的白兰地，将其陈酿调配后装瓶。作为一家年轻的公司，它积极尝试用陈年白兰地调制新式鸡尾酒。以下是一份用赫雷斯白兰地调制"可冷士"的配方。

50毫升赫雷斯白兰地

30毫升鲜柠檬汁

20毫升单糖浆

冰镇苏打水

柠檬片

将除苏打水之外的所有配料同冰块混合，倒入调酒壶调和，滤入加满冰的可冷士杯，再向杯中注满苏打水，用柠檬片装饰，最后插上吸管即可供人饮用。

白兰地库斯塔

这是弗朗索瓦普瑞米尔酒店路易丝酒吧的调酒师亚历山大·朗贝尔为我制作的一份配方，该酒店于2012年在干邑市中心开业。这款鸡尾酒的原版诞生于19世纪中叶，由约瑟夫·圣蒂尼在新奥尔良市的南方之珠酒吧首创。朗贝尔说，轩尼诗特选干邑是介于 VS 和 VSOP 之间的一款酒，他也推荐用 VSOP 来调配这款饮品。

1 抖振①北秀德苦精

1 抖振安高天娜苦精

1 茶匙原始配方橙皮甜酒

1 茶匙路萨朵樱桃利口酒

40 毫升轩尼诗"特选干邑"

2 茶匙鲜柠檬汁

1 茶匙单糖浆

白砂糖

柠檬皮

在玻璃罐中搅拌冰块，兑入液体配料，继续调和直至配料相互融合，滤入杯口抹好糖边的碟形杯，最后放入一条螺旋形薄柠檬皮。

干邑潘趣酒

英国人用白兰地调制出了潘趣酒。"潘趣"一词来自印地语，原意是"5"，意思是调制这种酒要用到 5 种配料：白糖、白兰地、柠檬汁或酸橙汁、水和调味料。有时会用葡萄酒来代替水，有时也会将葡萄酒和水一起加入。这个配

---

① 调酒的特殊计量单位，最常用于添加苦精时，指甩一下瓶子滴出来的量，大约是 1 毫升。

方来自法国干邑行业管理局。

4 只柠檬取皮

250 克（9 盎司）特细糖粉

250 毫升（1 杯）过滤好的鲜柠檬汁

750 毫升（1 瓶或 3 杯）VSOP（或 VS）干邑

250 毫升（1 杯）朗姆酒

1.5 升（6 杯）冷水

1 整粒肉豆蔻

将柠檬皮和糖粉混合拌匀，静置至少 1 小时。再次搅拌并加入柠檬汁，搅动至糖粉溶解。将柠檬皮滤掉，加入干邑和朗姆酒调和后放入冰箱冷藏。

供人饮用前，将冷藏液体倒入一只盛有一半冰块的潘趣酒碗中，加冷水调和，最后将肉豆蔻磨碎一半撒到上面。

干邑长饮

夏天到访干邑的游客，看到这里的人居然用苏打水或汤力水来稀释珍贵的干邑，总是会大吃一惊。不过，只要一杯在手，他们立刻就能领会个中妙趣。在干邑地区那些最炎热的日日夜夜，干邑长饮是永远的时尚。做一杯简单的"长饮"，通常用 VS 就可以了，不过如果能用 VSOP 当然更好。这个配方来自法国干邑行业管理局。

30 毫升（1 液量盎司）干邑

90 毫升（3 液量盎司）汤力水

将所有配料倒入装有冰块的可冷士杯，轻轻搅拌。

法国贩毒网①

这是一款适合晚餐后饮用的鸡尾酒，以苦杏酒的显著香气为干邑提味。亦可根据个人口味调整配比，将苦杏酒和干邑按照 1:2 的比例混合。

30 毫升（1 液量盎司）干邑

30 毫升（1 液量盎司）苦杏利口酒

将所有配料倒入装有冰块的海波杯中，搅拌至融合调匀。

马 颈

每年 7 月，干邑镇中心的一处公园里都会举办一场美妙的户外音乐节，名为"干邑蓝调激情"。值此温暖夏夜，人们会一边畅游音乐之海一边啜饮这款经典饮品。以下是法国干邑行业管理局提供的配方。

---

① 得名于 1971 年在美国上映的同名犯罪电影，但据说这款酒的配方早在 20 世纪 60 年代就已存在。

30 毫升（1 液量盎司）VS 或 VSOP 干邑

1 抖振安高天娜苦精

姜汁汽水

橙皮（亦可不加）

可冷士杯中放入几个冰块，倒入干邑，加 1 抖振苦精后再注满姜汁汽水。可取一长条螺旋形薄橙皮作为装饰。

热托蒂

其实这款酒就是以蜂蜜增甜，再兑入一"药用"剂量白兰地的热柠檬水。如果想要风味更足一些，可以用茶来代替热水。无论是加茶还是加热水，人们饮用这种酒预防受凉感冒已有数世纪之久。

1 汤匙蜂蜜

2 茶匙柠檬汁

125 毫升（半杯）滚水

30 毫升（1 液量盎司）白兰地

将蜂蜜、柠檬汁和热水倒入马克杯中，边搅拌边加入白兰地，搅拌完全后即刻享用。

药用白兰地

这款酒适合在感冒或喉咙发痒时饮用。过去人们认为，身体虚弱或病后初愈的人也不妨一试，因为牛奶能提供营养。这是意大利和其他许多国家的传统药方。

180 毫升（6 液量盎司）牛奶

30 毫升（1 液量盎司）白兰地

1 茶匙白糖（可选）

将牛奶加热至略低于沸点，也可以用意式咖啡机或奶泡器使牛奶起泡，之后倒入玻璃杯或马克杯，边加入白兰地边搅拌。如喜欢甜味可以加糖。早晨或晚间服用。

现代莫吉托

此配方由西班牙的赫雷斯白兰地制造商派卓多美酒庄提供。这是传统赫雷斯白兰地的一种非正式用法，是酒庄某位员工的最爱。各种材料的配比并没有具体的定量，这份配方只是酒庄的人在尝试过各种各样的配料组合方式后确定下来的。它本身就是一款既新鲜又健康的饮品，所以请尽管多加些薄荷叶和柠檬汁。

新鲜薄荷叶

2 茶匙白糖

1 汤匙鲜柠檬汁

50 毫升赫雷斯白兰地

碎冰

将薄荷叶和白糖放入调酒壶搅混，再加入柠檬汁和白兰地，充分摇和至白糖溶解，最后倒入装有半杯碎冰的曼哈顿杯。

皮斯科潘趣酒

这款酒的原始配方由当代专业调酒师设计，内含由取自金合欢树的阿拉伯树胶制成的糖浆。将阿拉伯树胶与糖溶液混合的过程十分费时，但调酒师们打包票说这等待绝对值得。以下简化版配方由秘鲁的波顿皮斯科酒庄提供，在家中宴客时制作起来比较方便。

1 只新鲜菠萝

235 毫升（1 杯）单糖浆

470 毫升（2 杯）瓶装水

750 毫升（1 瓶，或 3 杯）波顿皮斯科

300 毫升（1¼ 杯）鲜柠檬汁

取新鲜菠萝一只，切成约 0.5 英寸 ×1.5 英寸（1.5 厘米 ×4 厘米）的小块，放入单糖浆中腌渍一夜。次日早上，将其余配料放入一只大碗混合，可加入柠檬汁或单糖浆调味。每只玻璃杯中倒入 3 至 4 液量盎司（约 100 毫升）潘趣酒，再加入一块腌渍好的菠萝。

皮斯科酸酒

秘鲁人开派对总是少不了皮斯科酸酒。以下配方由创立马丘皮斯科品牌的两姐妹提供。

2 份秘鲁皮斯科（最好是酷斑妲葡萄酿制的皮斯科）

1 份鲜酸橙汁

1 份白糖

1 子弹杯[①]蛋清

安高天娜苦精

冰

电动搅拌器中加两杯冰，再将前四种配料放入，充分搅和后倒入冰镇过的玻璃杯。每杯鸡尾酒中滴上几滴安高天娜苦精。做几份呢？你说了算……

边　车

这份配方来自法国干邑行业管理局，其渊源要从调酒师达勒·德格罗夫讲述的这段边车辉煌历史说起：

边车的配方同白兰地库斯塔一脉相承，不过在这种酒的相关记载中，20 世纪的篇章却是语焉不详。哈利的纽约酒吧声称边车是它的首创。然而，巴黎丽兹酒店海明威酒吧的首席调酒师科林·菲尔德确信，这款饮品其实是他的

----

① 　大约 30 毫升。

前辈——海明威酒吧早年间的传奇调酒师弗兰克·梅耶尔于 1923 年发明的，但这种说法并无相关材料佐证。我们能找到的唯一书面证据，就是伦敦使馆酒吧的罗伯特·韦梅尔写于 1922 年的一本书——《如何调制鸡尾酒》。书中说，边车的创制者是伦敦巴克俱乐部的一位调酒师，名叫麦克加里。

45 毫升（1½ 液量盎司）VSOP干邑

30 毫升（1 液量盎司）橙皮甜酒

20 毫升（¾ 液量盎司）鲜柠檬汁

橙皮（亦可不加）

将所有配料倒入调酒杯中混合，滤入挂少许糖边的小型鸡尾酒杯，饰以一小条火烧橙皮。

史汀格

这款鸡尾酒曾是多年的经典，不过今天它或许不再是所有人的首选了。酒名大概就是指你在喝下这款饮品时所体验到的那种"刺激感"[①]。薄荷酒的用量可以根据个人喜好适当减少，比如有些人喜欢以 2:1 的比例来搭配白兰地和薄荷酒。

---

① 英文 stinger 的意思是螫针，因此这款酒有时也被译为"毒刺"。

30 毫升（1 液量盎司）白兰地

30 毫升（1 液量盎司）薄荷酒

碎冰

鲜薄荷叶（亦可不加）

将配料倒入盛有碎冰的调酒壶摇匀，滤入鸡尾酒杯。可饰以鲜薄荷叶。

峰会鸡尾酒

几年前，法国干邑行业管理局开始鼓励调酒师尝试用干邑调制鸡尾酒。为此，管理局举办了一场鸡尾酒峰会，邀请许多知名调酒师参加。这款鸡尾酒就是安迪·西摩专为此次峰会而创作的，现已成为新晋经典，在世界各地许多干邑管理局的活动上都有供应。

一只酸橙取皮

4 薄片鲜姜

45 毫升（1½ 液量盎司）VSOP 干邑

60 毫升（2 液量盎司）柠檬酸橙苏打水

1 长片黄瓜

将酸橙和姜放入玻璃杯中，倒入一半干邑，轻轻按压 2 至 3 次。加冰至半满，搅拌 5 秒钟。倒入剩余的干邑，加入柠檬酸橙苏打水及黄瓜，充分搅拌后即可供人饮用。

至尊雅文邑

这是巴黎克利翁酒店首席调酒师菲利普·奥利维耶与法国雅文邑行业管理局最近共同创作的一款鸡尾酒。它将柑橘与白兰地混合调制，这种经典风味的搭配组合能满足世界各地诸多人士的口味需求。

40 毫升（1¼ 液量盎司）雅文邑

30 毫升（1 液量盎司）葡萄柚汁

2 茶匙用去皮橙瓣榨出的橙汁

马拉斯奇诺樱桃[①]（亦可不加）

将所有配料放入加冰的调酒壶中，充分摇和后滤入鸡尾酒杯。可用马拉斯奇诺樱桃装饰杯沿。

[①] 得名于其传统制作工艺：将马拉斯加酸樱桃的籽、梗、叶子和果肉蒸馏成名为马拉斯奇诺的烈酒，再将樱桃浸在酒中保存。

致谢

亚美尼亚：亚拉拉特公司／保乐力加集团，诺伊公司，普罗尚白兰地公司，韦迪阿尔科公司；雅文邑：阿曼达·加纳姆，法国雅文邑行业管理局及其所有成员，塔西克酒庄的伊捷·布沙尔，让·卡斯塔雷德，朗巴德酒庄的阿尔诺·莱古尔格和德尼·莱古尔格；澳大利亚：安戈瓦公司的马特·雷丁，精品葡萄酒拍档公司的罗伯·赫斯特，塔南达酒庄的约翰·格伯和马蒂·鲍威尔；干邑：让－路易·卡尔波尼埃，尼基·塞兹摩尔，法国干邑行业管理局及其所有成员；格鲁吉亚：格鲁吉亚葡萄酒协会的蒂娜·克泽利和乔吉·阿普哈扎娃，叶卡捷琳娜·埃古蒂亚，兹维亚德·克利维泽，大卫·阿布齐亚尼泽和萨拉吉利公司，第比利斯马拉尼公司，卡赫季传统酿酒公司；赫雷斯：贝亚姆·多梅克，赫雷斯白兰地监管委员会的卡门·奥梅斯克和塞萨尔·萨尔达尼亚，卡斯蒂利亚费尔南德酒庄，冈萨雷斯·拜亚斯，桑切斯·罗马德，传统酒庄的洛伦索·加西亚－伊格莱西亚斯；皮斯科白兰地：马丘皮斯科公司的伊丽莎白和梅兰妮·亚瑟，波顿皮斯科酒庄的约翰尼·舒勒，吉列尔·L.托罗－里拉；南非：路易斯·德科克和泰莎·德科克，艾尔莎·福格茨，凯樽汇酒业集团；美国：美国烈酒研究所的比尔·欧文斯，嘉璐酒庄的斯科特·迪萨尔沃和拉塞尔·里基茨，手指湖酿酒厂的布莱恩·麦肯齐，安斯莉·科尔，休伯特·热尔曼－罗班，奥索佳公司的丹·法伯。

　　另外还要感谢：大卫·贝克，蒂姆·克拉克，康科德作家团，吉尔·德格罗夫和戴尔·德格罗夫，皮耶路易吉·多尼尼，布兰科·格罗瓦茨，凯尔·贾拉德，人头马美国公司的劳伦·基内尔斯基，桑德拉·麦克唐纳，查理斯·德伯纳，马尼埃－拉博丝特公司，马斯卡罗公司的玛丽亚·马塔，伊丽莎白·明基利，比姆全球公司的马克·帕洛，诺姆·罗比，肯恩·西蒙森，哈米什·史密斯，贾恩·所罗门，卡尔文·斯托瓦尔，鸡尾酒传奇基金会的安·图纳曼，布兰卡兄弟酒厂的艾丽莎·维努达和科尔斯顿·阿曼，罗西·维达尔，大卫·瓦德里希。

Brandy

A Global History

Becky Sue Epstein

Contents

Introduction: Luxury Cognacs, Alluring Brandies   145

Chapter 1   Alchemy: From Classical
             Civilizations to Cognac   151

Chapter 2   Producing Brandy: Distilling and Ageing   157

Chapter 3   Cognac becomes World-famous   164

Chapter 4   Armagnac and its Noble History   174

Chapter 5   Illustrious Brandies of Europe and
             the Caucasus   180

Chapter 6   Great Spanish and Latin Brandies   189

Chapter 7   Australia and South Africa   201

Chapter 8   Brandy Made in America   208

Chapter 9   Everything You Need to Know
             about Cognac   214

Chapter 10   Cognac Cocktails and 21st-century Trends   224

Chapter 11   Small-batch Brandies and Cognacs   231

Recipes   238

Select Bibliography   250

Websites and Associations   252

Acknowledgements   256

Photo Acknowledgements   258

Introduction:
Luxury Cognacs, Alluring Brandies

No, sir, claret is the liquor for boys; port for men; but
he who aspires to be a hero (smiling), must drink brandy.
In the first place, the flavour of brandy is most grateful to
the palate; and then brandy will do soonest for a man what
drinking can do for him.

James Boswell, *The Life of Samuel Johnson* (1791)

Today, brandy is back in the limelight. It is cocktail chic. It is a luxury sipping spirit in thousand-dollar crystal decanters. It is the party drink of celebrities. Yet only a few decades ago, brandy was far removed from this illustrious position.

Everyone knows what brandy is—or do they? Brandy is a wonderful, aromatic spirit made by distilling wine. It is usually amber- or mahogany-coloured as a result of its ageing in wooden barrels. Today, the most famous brandies in the United States and Britain are cognac and armagnac, which are produced in two different regions of southern France. Spanish and other brandies are also popular in various parts of the world.

Until a few decades ago, expensive brandies were carefully measured out by the wealthy as after-dinner drinks. Older people might keep a bottle of basic brandy at the back of a cupboard, for vaguely medicinal purposes. A certain amount of brandy was sold as cheap tipple. But for most of us, brandy simply wasn't part of our lives.

One bright spot remained: cognac, the most famous brandy in the world. Brandies from Cognac have always retained their cachet. When asked, most people have a favourable view of cognac even if they've never encountered it. Cognac is the most famous and expensive brandy, a concept reinforced again and again by today's stars. Legendary film director Martin Scorsese has been featured in adverts for Hennessy. The rap music star Ludacris has his own brand, Conjure; Snoop Dogg has endorsed Landy Cognac, and many other successful rappers mention favourite cognacs in their hit songs. In the current high-end bar scene, stylish bartenders (now called mixologists) compete in cognac cocktail contests, selecting subtle flavourings to create exceptional drink blends that enhance this elegant, aromatic spirit.

Technically, brandies can be distilled from a variety of fruits, but for the purpose of this book, brandy is defined as a spirit made from grape wine. The wine is distilled into a spirit, and then most often aged in wood, which gives brandy its lovely tawny colour. This is what we think of as 'brandy'.

While young brandies can inspire cocktails and

relaxation, sipping an aged cognac or brandy is a truly memorable event. We experience a wonderful sense of ease and well-being when a fine spirit of 40 per cent alcohol courses through our bodies. (Medicinal? Perhaps.)

As late as the end of the twentieth century, when brandy's fortunes were in decline, it was cognac that sparked its resurgence. Until that time, upmarket brandy from Cognac was seen as a drink for prosperous older people—mainly men. Through the middle of the century, it was only considered acceptable for women to consume cognac in public in dishes like crêpes Suzette, which is dramatically flamed at the table with a brandy-based sauce. Around the world, less wealthy populations with brandy traditions sipped much less expensive brandies, most often lower-quality domestic brands made from grape wine (as well as other substances) but still called 'brandy'—or even 'cognac'. These highs and lows were continuations of brandy's long progression through civilization, a journey that began more than 700 years ago.

Beginning in the Middle Ages, brandy was prescribed medicinally for a vast number of ills and conditions. In the modern era, brandy also served as the basic ingredient for many of the punch drinks that were extremely popular in the eighteenth and nineteenth centuries in the United States and Britain. During the middle of the nineteenth century, brandy became the base spirit for the first-ever cocktail craze, which originated in the U.S .

Due to its exacting distillation process, brandy is not an inexpensive spirit to produce. Factor in years of barrel-ageing, and still more is added to the price of a bottle. Cognac, armagnac and brandy de Jerez are the three most famous—and expensive—noble Old World brandies. Early in their histories, South Africa and Australia established thriving brandy industries because they were colonial outposts of Britain and the Netherlands, where people traditionally consumed cognac. High-quality, grape-wine-based brandies have been made in Armenia, Georgia, the U.S. and other countries for over 100 years.

Since cognac was the most widely known luxury brandy, many countries modelled their production on that of the Cognac region. Most countries even went so far as to call their brandy 'cognac'. This gradually became an identity problem for France's cognac producers, an issue that they have been actively dealing with since brandy production surged across the globe in the late 1800s.

Also towards the end of the nineteenth century, the availability of fine cognac from France began to decrease due to the phylloxera epidemic that decimated European vineyards. By the beginning of the twentieth century, with cognac's high prices and constrained availability, whisky and other spirits replaced brandy in many fashionable cocktails. So brandy—and cognac in particular—became even further removed from the daily world of most consumers. This situation continued through the middle decades of the century, when countless scenes in books and films reinforced the image of snobbery associated with drinking brandy. The suave spy James Bond created a little stir in the brandy world in 1971; in the film *Diamonds are Forever* he briefly took cognac out of the drawing room and into the line of fire, literally, when he used Courvoisier to flambé an opponent. But all in all, it appeared that brandies were being relegated to the increasingly rarely used drawing room.

Yet suddenly, at the turn of the twenty-first century, brandy—more specifically cognac—was dusted off and brought into the mainstream, largely through the genre known as urban music and the unlikely agency of rap and hip hop artists. During the latter decades of the twentieth century, lower-end brandies had been very popular in certain urban areas in the U.S. When urban music styles took off, even young people who didn't live in these areas began to emulate the lifestyle they heard on the radio and saw in videos. As the number of hip hop and rap stars increased, more mentions of luxury bottles of Courvoisier, Hennessy and other top brands began to show up in their songs and in their music videos. And fans followed enthusiastically into the world of luxury cognac. Abruptly, cognac consumption exploded.

At the same time, bartenders were also becoming more creative, turning into professional 'mixologists'; and cocktails were gaining in popularity in upmarket urban establishments. Though it may be impossible to proclaim either music or mixologists as the ultimate catalyst, the result is clear: brandy—and cognac in particular—is in the ascendancy again.

Today, in New York and other cosmopolitan cities, people are more likely to have their brandy before dinner than after. In the early evening behind the bar, cognac is often poured into a cocktail shaker. There, mixologists perform their magic by adding select flavourings and other liquors and liqueurs to turn it into the perfect, chilled cocktail. Increasingly, adventurous imbibers sample brandy for the first time this way: in a heady concoction at an upscale bar. This is especially true of cognac. Expensive cognac cocktails have taken off in major U.S. cities, and other urban areas around the world are starting to follow this enticing lead, guided by internationally oriented mixologists.

For aged brandies, the UK and the U.S. have been the traditional leaders in consumption. But China is currently overtaking the West, after more than a decade of soaring consumption that shows no sign of abating. And there's still a huge market for basic brandies around the world; Malaysia, the Philippines, India and other Asian countries are all major brandy consumers. How can this spirit be popular on so many levels? And what are all these levels?

In addition to place of origin, ageing makes the most difference to the price and quality of brandies. But they all start with the same process: distillation. It has been a long journey, one that took centuries, as distillation moved from the centre of the ancient world to Europe and then to the New World. This is the story we will explore.

Chapter 1
Alchemy: From Classical Civilizations to Cognac

Fire and gold, Dionysus and Cleopatra, early Christianity and secret sects all played a part in the history of distillation. Early distillers in classical civilizations had several different goals. Some were looking for an elixir of life, or *aqua vitae*. Others were also producing *aqua ardens*, the miraculous combination of two opposing elements, water and fire; *aqua ardens* was a magical 'burning water', an ignitable liquid. Later, in the early Middle Ages, alchemists created different types of distillates while seeking to create the highest of all metals, gold. Early Egyptians explored distillation as early as Cleopatra's reign in Egypt in the first century BC, when two slightly different types of distilling were practised by respected philosopher-chemists, and distilled wine was also used by Greek followers of the cult of Dionysus.

Though distilling seemed to have disappeared once the Roman Empire became Christian, in reality it only went underground. For centuries, the process was used to prepare fluids for the sacred rituals of secret Gnostic sects. During the course of a thousand years, religious Cathars underwent a true 'baptism by fire' involving 'burning water' (*aqua ardens*) which was created by distillation. The technique may also have spread to Asia—or developed simultaneously. Early Arab alchemists had apparently heard rumours that distillation was used by Taoists in China to create an 'elixir of life' (*aqua vitae*) in the fourth century. But in the West in the third century, even common sailors understood the concept of distilling; essentially they evaporated sea water and then carefully condensed it to obtain the desalinated water they desperately needed on long voyages. To them, this was the 'water of life', the *aqua vitae* that literally kept them alive.

It was only one more step to use the concept of distilling as a weapon. Warring sailors used *aqua ardens* at the Battle of Cyzicus, an island in the Sea of Marmara, around the year 672. There, Byzantines successfully defended the great city of Constantinople by throwing flaming liquid at the attacking Saracen ships. Though this 'Greek fire' may have involved petroleum in addition to (possibly instead of) distilled wine, it nevertheless became part of the lore of distillation's power, and the force of its resultant product, *aqua ardens*.

Knowledge of distillation also spread through the Middle East to the Persian Empire, where it was used to achieve more positive results in the emerging science of herbal medicine. In the sixth century, the Persian shah Khosrau I established a medical school in his city of Gundeshapur. Surrounding the academy were gardens filled with herbs, flowers and other botanical specimens. Distilling alcohol was critical here, as it was used to prepare all kinds of medicinal infusions.

Under the Moorish occupation, this non-drinking culture did not destroy Spain's existing vineyards, because grapes were also necessary for the distillation process to make perfumes and make-up. In fact, the word 'alcohol' is thought to come from the Arab word for the dark eyeliner known as kohl (*al* means 'the' or 'a' in Arabic, forming *al-kuhl*). The name 'alembic' for the type of still also came from Arabic, which in turn was derived from *ambix*, the Greek word for cup. The word 'alembic' is found in written French as far back as 1265.

In Europe, distilling was part of the process alchemists used to try to turn ordinary items into gold. In fact, one chemical process involving distillation gives certain items a gold-coloured coating; this seems to have been enough to encourage many more generations to try to develop the process to make real gold. In addition to being considered the most important metal, gold was believed to be a cure for various illnesses and conditions. (The modern German liqueur Goldwasser, 'gold water', is a reminder of this belief.)

As chemistry, alchemy and medicine became intertwined during the Middle Ages, distillation became part of all of these fields. By the late Middle Ages, distilling for medicinal preparations had spread throughout northern Europe and the British Isles, through alchemists and physicians. But the world was now ready for another use—shall we call it recreational?

However, the results of distilling wine were mixed in the early days. For instance, in order to sell more wine, German merchants distilled wine with additives that included poisonous chemicals. Many people attempted to distil matter other than grape wine, which had a variety of results, mostly bad. Eventually, this experimentation led to the production of whisky, which of course was good. But before that, we had brandy.

Drinkable brandy was reliably produced by distilling grape wine. One of the first commercial uses of potable brandy—at the time a clear, un-aged spirit—was to fortify existing wine. A spirit can be added to wine to stabilize it against spoilage by strengthening its alcohol content. Sometimes, to create a sweeter wine, the spirit is added during fermentation, which kills the yeast before it can convert all the grape sugars into alcohol. The great sweet wines of Roussillon from the Mediterranean coast of France are still fortified with locally distilled spirits today.

The science of distilling had moved further north into the Gascony area of France by the thirteenth century, perhaps brought by pilgrims returning from Santiago de Compostela. And in 1299 , we have evidence that Arnaud de Villeneuve, the private physician of Pope Clement v, was treating him with medicine made from distilled grape wine. He called it the water of life, *aqua vitae* in Latin; in French it was *eau de vie*, which remains the common French term for a distilled spirit to this day.

A short time later, in Armagnac (located in Gascony)

we find this wine-grape distillate being stored (aged) in barrels, thus launching the modern era of brandy. It was the Armagnac region that gave its name to the first aged brandies in 1310, just over 700 years ago. From Gascony, brandy distillation spread further: south to the Andalusian region of Spain where the town of Jerez is located—Jerez is not only the home of sherry but also of the great brandy de Jerez— and north along the Atlantic coast of France, to Bordeaux, Cognac and the Loire Valley.

Dutch traders sailed up and down the Atlantic coast of Europe from the early sixteenth century. In particular, they brought home wines from France because it was too cold to grow wine grapes in Holland. Contrary to what is written in most short histories of brandy, there was no instant, meteoric rise in the dissemination of distilled spirits from Cognac. In fact, Dutch traders took their wines from towns along the rivers that fed into the Atlantic coast of western France. Perhaps initially Dutch traders distilled some of the weaker wines in order to stabilize them for the journey home, or perhaps it was simply more efficient to ship concentrated, distilled wine; this was first called *brandewijn* (cooked or 'burned' wine), a term which was later shortened to *brandy*.

The Dutch drank brandy from at least the year 1536; this is documented in a regulation which prohibited tavern-keepers from selling brandy to be consumed off their premises. Britain imported brandy for householders to make cordials, to 'strengthen' weak wines and to provide a base for medicinal herbs and spices. But British, French and Dutch business dealings in cognac were severely interrupted by the Glorious Revolution and the Nine Years' War at the end of the seventeenth century. Of course, a brisk trade immediately grew up in counterfeit brandy—distillations made with all sorts of fruits and spices that attempted to mimic the French drink's flavours.

Smuggling was another 'industry' that benefited from the politics of the seventeenth and eighteenth centuries. Even at times when there was no ongoing war, there was a plentiful market for brandies that had evaded British taxes by being put ashore on the many inlets along the English coastline. The Irish gentry drank brandy, too, and they even sent some of their merchants to Cognac to run operations there—which is why some of the most famous cognac houses today have Irish names, like Hennessy and Otard (originally O'Tard).

In Europe, Britain and America, brandy was considered medicinal, and in most households it was kept as a remedy for everything from fainting to indigestion. Travellers carried it for energy and as an antiseptic. 'Medicinal brandy' is a phrase many people today have heard—or used—without knowing exactly what it meant. Whether considered a necessity of life or a recreational beverage, the distilled grape wine called brandy has been a staple commodity throughout the Western world for the past several centuries.

Chapter 2
Producing Brandy: Distilling and Ageing

Everyone knows brandy is a golden-brown spirit. Or is it? Most consumers—even connoisseurs—are not familiar with brandy's method of production. In fact, brandy starts out clear. Most brandies are aged in wood, which imparts a lovely amber hue to the clear liquid that deepens over time. But whether clear or coloured, the spirit is still brandy. In fact, one of the newly popular spirits in Western mixology is a clear brandy called pisco. Yes, pisco, which originated in Peru, is a brandy. It's made in the same way as cognac from France and brandy de Jerez from Spain: distilled from grape wine. But it isn't generally aged in oak barrels, hence its clear colour.

Grape-wine-based brandy, which is the focus of this book, is one of three main types of brandy. The others are distilled from other fruits and from pomace. Well-known brandies made from fruit include calvados (apples), slivovitz (plums) and kirsch(cherries), but there are many more national and local favourites around the globe. Pomace-based brandies are distilled from the skins, stems and seeds of wine grapes—the leftovers after the grapes have been pressed for wine. Two of the most famous pomace brandies are grappa from Italy and marc, a name that originated in France. Why are we not talking about fruit and pomace brandies in this book? Because most people don't think of spirits like grappa as brandies. And there are so many amazing grape-based brandies around the world.

What all three categories of brandy have in common is that they are made by distillation. Distillation is what turns them from their raw material—grapes, fruit, pomace—into a spirit. Recalling the history of distillation, it was a process used by alchemists in the Middle Ages before brandies started to become well-known, tradable commodities. Perhaps it's no accident that the word 'spirit' refers to a substance or essence that is vital and strong, both in the context of alcohol and in the disciplines of philosophy and religion. Having discovered earlier how distilling made its way from the Middle East into Europe, it will be interesting to see how this process is used in the production of brandy in various regions.

Very simply, distillation is the process of heating wine in a closed container, then capturing the resultant aromatic, alcoholic vapour, and cooling it into a liquid. The wine is heated to its boiling point, a temperature at which more alcohol than water evaporates. (Alcohol boils at 78°C (172°F), while water's boiling temperature is 100°C (212°F).) The alcohol is concentrated in the resulting condensed liquid: this is when wine becomes a spirit. And some of the wine's aromas and flavours are carried in the alcohol as it evaporates and then condenses, which gives each brandy its own distinct character.

Of course, brandy distilling is not quite as simple as boiling up a wine; if it was, anyone would be able to do it. In reality, it can be quite a complex, beautiful process, with a bright copper vessel shining like a giant kettle in a large room redolent of wood smoke from the fires underneath. The huge copper pot is connected to other containers with angled, columnar or even corkscrew-shaped pipes, so that the whole collection looks like a massive, playfully gleaming laboratory apparatus—especially pleasing when you consider the wonderful product being created.

Three main factors influence the creation of brandy: the type of wine, the type of distilling equipment and the skill of the distiller. Each brandy region also has its own set of regulations and traditions that contribute to the final product. These include elements such as the type of still that may be used, which months of the year distilling must take place in, the ageing process and the shapes of the bottles.

Most of the wines used to make brandy are made from white grapes that have two properties. First, they have good acidic components. Second, the grapes are capable of imparting desirable aromas and flavours to a brandy after distillation. Traditionally, grapes used for brandy did not make great wines—though that is not universally true.

After the grapes are harvested, they are fermented into wine. It is important that the wine is distilled immediately (unless it can be held in a temperature-controlled environment) because the common wine preservative sulphur dioxide (SO_2)

would effectively ruin the distilled spirit. Historically, this is why Cognac and Armagnac dictated that wine distilling had to start on a certain day (just after the grapes had been fully fermented into wine) and finish by the end of the winter, when the weather was still cold enough to preserve the wine naturally. Nowadays, wines can be held in chilled tanks for weeks or months until they are ready to be distilled, but the tradition endures in the rules set by the regions.

Most commonly, wine is taken off the lees—the dregs of the yeast and other residue in the wine that settles during fermentation—before it is distilled, as it is easy for the lees to fall to the bottom of the still and burn, ruining a batch of brandy. However, those producers who do distil with the lees say it makes a more flavourful drink, worth the extra effort of installing a propeller-type mixer in the still to keep the lees circulating in the wine.

The distillation process of an entire vintage of wine can take several months, since there is generally a limited number of stills—and expert distillers—available at any given facility. In the world of brandies, there are basically two types of distilling process: single distillation and double distillation. Because multiple distillations are currently being popularized in the vodka world, you might think that more is better. In brandy, it's not. Ask anyone from Armagnac, where single distillation is the honoured tradition. But move to Cognac, and you might think that double distillation is the only way to go.

Both distillation processes produce many impurities,

from foul aromas or terrible flavours like fusel oil (which tastes more like something you'd use in an engine than a drink) to downright dangerous chemicals. 'Heads' are compounds that are more volatile than ethanol, meaning that they evaporate and condense at a lower temperature than ethanol; 'tails' are compounds less volatile than ethanol, and evaporate and condense at higher temperatures. During double distillation, heads and tails are typically discarded at the beginning and end of the second distillation. Double distillation is practised in Cognac as well as in some other regions. The wine literally goes through the process of distilling twice; sometimes parts of the heads or tails are added into the second distillation, sometimes not, depending on the company's tradition and the master distiller's style.

Devotees of the single-distillation process—such as those in Armagnac and Jerez—will tell you that this is a far more exacting process than double distilling. In single distillation, the temperature gradient of the distillation column separates the heads and tails from the desirable spirit. The master distiller must carefully manage the temperature and operation of the column still in order to remove the undesirable compounds and to keep all of the desirable aromas and flavours and the degree of alcohol in a single process.

Many distillers use gas heat, but some distilleries use wood fires to heat their stills. At Château du Tariquet in Armagnac—as well as a few others in this region, and in Cognac and Jerez—the smoky aromas greatly enhance the experience for lucky visitors during the autumn distillation period.

After distillation, the finished spirit is ready to be aged. The wood-ageing process contributes enticing aromas ranging from vanilla and toffee to dried fruits, cedar and spices. In most regions, towards the end of the ageing period, a certain amount of carefully filtered water must usually be added to dilute the brandy to its final drinking strength, which is around 40 per cent alcohol. 'Cask strength' (the alcohol content straight out of the barrel) is for most brandies around 60 to 70 per cent—though this level changes over the years as both alcohol and water evaporate (at different rates) during the ageing process.

Every region has a preferred wood that it uses to make the barrels for ageing its brandies. Limousin oak, from near Cognac, is considered the gold standard, but there are other forests of oaks with a similar grain in Armagnac, the Caucasus Mountains and elsewhere. Some brandy producers actually go to the forest to choose the trees to use for barrels, while others may select already cut wood at their cooperage. Wood for the ageing barrels must be left out to air-dry for several years.

Most of the barrels for cognac and other high-end brandies are handmade. First the dried staves are heated gently to make them flexible, and then they are assembled like a giant upside-down flower around a small fire. In Cognac, this is called *mise en rose*, or putting it into a rose-shape. Bands are then placed on top and slowly hammered down into place, holding the staves together. After a barrel's inside is 'toasted' to order, the ends are added, and the finished barrel is rolled out of the hot, smoky work area, as has been done for hundreds of years.

Barrel-ageing rooms are dark and quiet and may be above ground or below. As brandies develop, they may be moved from warmer to cooler, or from more to less humid areas. Originally, small producers kept their barrels in their farm outbuildings, and many still do today, while large cognac houses age barrels in their cellars or in other types of warehouses.

But in Cognac, all the ageing rooms have one thing in common: spiders. It is bad luck to kill the spiders. Until very recently, cognac was aged in barrels held together with willow bands, because willow is flexible. However, there is a certain type of mite in Cognac that likes to eat willow wood. The spiders are there to eat the mites, thus keeping the cognac barrels intact for decades, so cognac producers take care not to disturb them. Even now, when most barrels are bound with metal bands, there is still a superstition that a healthy cellar must contain plenty of spiders—perhaps a far cry from the luxurious image of this spirit, yet essential.

Chapter 3
Cognac becomes World-famous

It is impossible to tour the small art museum in the city of Cognac without noticing that the man who was king of France 500 years ago—Francis I—tends to be smiling in his portraits and statues. And he had every reason to be happy. During his reign, the riverside town of Cognac was already well known as a shipping centre in the region, and the brandy business was just beginning. Dutch merchants were expanding their reach in the area, looking for new products to trade.

The most pervasive theory about the 'invention' of cognac has to do with Dutch traders fortifying the local wine for shipping, only to reconstitute it with water on arrival in their home cities. But this is a simplification of what really happened. Dutch merchants were the savviest traders of their era. They had a single-mindedness of vision: they knew what to buy and what to sell, what to build and how to do it, in order to maximize their profit. They traded tirelessly along the Atlantic coast of France from the sixteenth century on. One of the more important commodities they brought to the Netherlands was the salt from the French coastal trading port of La Rochelle. La Rochelle sits just across a narrow channel from the Île de Ré–which is still famous today for its excellent-quality salt. The city also happens to be located very near the mouth of the Charente, the river that runs through Cognac. And, of course, there was the wine.

The winemakers of the Loire Valley shipped their wines to a merchant colony the Dutch had set up in the large trading city of Nantes on the Loire river, on the way to the Atlantic. The Dutch set up another colony further south, in the Charente river region. They also traded with Bordeaux, which was an easy trip along the Gironde estuary from the Atlantic.

Because some of these French wines could be fairly light in flavour and chemically fragile, traders sought a way to stabilize them for transportation. They also wanted to improve the wines' ageing potential, though not in the way we think of today, of ageing bottles of wine for years until the flavours mature. In those days, wine had to age only well enough to last out the year without spoiling, until the next year's grapes were harvested and made into wine.

But the wines from the Loire and from Bordeaux proved to be more valuable when sold as wines than as distilled spirits. So while the Dutch continued to import and sell the wines from the Loire and Bordeaux, they kept on distilling wines from the Cognac area because they were more profitable.

This was also a century of political and religious upheaval, which had its effects on the cognac trade. Wars and conflicts between the English and the French continued. Catholics and Protestants tangled in local, national and international clashes in France, England, Holland and other northern European countries.

One important local battle that occurred in the seventeenth century greatly impacted Cognac's rise to economic dominance in the region. During a battle in 1651 , the people of the walled city of Cognac succeeded in holding back forces hostile to the Bourbon prince Louis II , who was a general in Louis XIV's army. In gratitude, King Louis later exempted Cognac from taxes and duties on its wines and distilled wines. With this financial advantage, Cognac soon began to outstrip its neighbours in economic development, becoming a regional commercial centre for all locally exported products—which included its brandy. Cognac was on its way to becoming the most famous brandy in the world.

Additional European and British political strife continued into the eighteenth century. Some Englishmen came to the Charente area to establish their own cognac production houses, so as not to be dependent on the French for a favourite—and quite profitable—spirit. Periodic Anglo-French conflicts also allowed the Irish to get into the cognac industry.

As this happened, the Dutch merchants lost their near-monopoly on the French brandy trade. Dutch citizens continued to crave spirits, but by then brandy was not the only one available to them. Developments in distilling had allowed people who lived in northern countries to create spirits from grain in places like Holland where it was too cold to grow wine grapes.

All these spirits, including brandy, were initially referred to as *eaux de vie*. But they soon acquired their own designated names, taken either from the local language or place where they were made. The Dutch had genever (or jenever; an early version of gin); the Scottish were starting to make Scotch whisky; Russia and other northern European countries began to produce vodkas; and rum arrived from the New World. But at the higher levels of society, the people of all these countries still enjoyed their cognac—especially in Holland, Germany and England. French, English and Irish names began to appear on the fronts of cognac houses in the first half of the eighteenth century. Some of these are still the top cognac producers today—names like Martell, Hennessy and Rémy Martin.

As this was happening, cognac began to differentiate itself from the other northern European spirits (or *eaux de vie*). Yes, they were all ardent, life-enhancing spirits, descended from a combination of alchemical, medical and religious practices. But cognac was a bit smoother, a little more refined—and more flavourful than other *eaux de vie* that existed at that time.

Several more elements aligned to contribute to cognac's quality and fame. One was the proximity of the Limousin forests, the source of wood for excellent cognac-ageing barrels. Another was cognac's period of popularity in Paris— and subsequently in French colonies around the globe. By the early eighteenth century residents of the north of France had become more aware of cognac. Unlike the English, they did not limit their consumption to high-end sipping; instead, they began to drink basic cognacs during the years when weather extremes caused problems in the vineyards and made local wines scarce. In fact, cognac has always had less cachet in France than in other countries, partly because of this early use of cognac, and partly because cognac was mainly considered an export product.

In order to be exported, cognac had to be put into barrels, made, of course, from local wood. The forests around the cognac production area were well known even in the Middle Ages. Later, their wood was used by Louis XIV's navy. This Limousin oak is considered top-quality even today. In Limousin oak (English oak or French oak, *Quercus robur*), cognac ages perfectly; the relatively loose wood grain of these trees allows just the right amount of liquid penetration inside the barrels, which transfers the ideal amounts of colour, spicy aroma and flavour from barrel to spirit.

The amber hues of cognac aged in barrels further distinguished cognac from *eaux de vie* that were produced from other fruits and grains at the same time. Those other spirits tended to be consumed in their local areas, and without barrel ageing they were clear in colour. Soon, barrel ageing became such a defining feature of the spirits from Cognac (as in Armagnac) that producers mastered the art of barrel construction and management solely for the enhancement of their cognacs.

As barrel ageing became an important characteristic of cognac, the concept of ageing itself took on new meaning. This is how the term 'Napoleon cognac' developed. At present, Napoleon cognac has a specific ageing requirement in the laws of Cognac, but this specificity is an only relatively recent development in Cognac, and is not generally adhered to around the world.

The motto of the famous Courvoisier cognac house is 'Le Cognac de Napoleon'. Courvoisier began as a wine and spirits company on the outskirts of Paris, which Napoleon Bonaparte is said to have visited in 1810. Perhaps because of this, he began issuing morning cognac rations to hearten his troops during the long Napoleonic Wars. Courvoisier was proud to advertise its connection with the emperor. Some years later, in 1828, after deciding to concentrate on cognac, Courvoisier relocated its company headquarters to the Cognac region. To this day Courvoisier headquarters remain there, in the town of Jarnac, on the Charente river just upstream from Cognac.

Some claim that Napoleon's advocacy of cognac was the reason many more brandy producers (in Cognac and in other areas of the world) began to name their cognac after the emperor. Others believe that the status of an aged cognac became a more important factor as the nineteenth century progressed, so the term 'Napoleon cognac' was used by more producers to designate valuable cognac that had, allegedly, been aged since the time of Napoleon. In any case, the name Napoleon served to increase both the cachet and the sales of cognac—so much so that it was often appropriated for other brandies in other countries, which is a problem Cognac's producers are still dealing with today.

As the era of cognac's importance unfolded, in Britain cognac overcame a challenging protectionist tax when Prime Minister Robert Peel cut cognac tariffs by nearly a third in the 1840s. This was followed by a further tax reduction in 1860, and six years after that cognac shipments to Britain had doubled.

In Britain, it was common for some distributors to ship young cognacs in barrels direct from Cognac, and then age them in dockside warehouses. After bottling, the cognacs might be labelled with the name of the trusted merchant, and not that of the cognac producer. These cognacs were called 'early landed' and had a slightly different taste and aroma profile because they were aged in a climate different from Cognac's.

In America by that time brandy (usually cognac) had become extremely popular as the basis for cocktails. This was during the (first ever) cocktail craze, which lasted through most of the nineteenth century. In France, cognac also got its first bottling factory, so that customers could identify and appreciate the tawny colour of this unique spirit when it was shipped in glass bottles instead of barrels. Everything was going swimmingly for cognac until the end of the nineteenth century, when it all came crashing down.

In the United States, cognac had been the spirit of wealthy men throughout the country, notably in the South.

But after the Civil War (1861–5) much of the Southern economy was ruined, and that market for cognac dried up. Americans had also begun producing their own whiskey, and the country developed a taste for its native spirits such as bourbon and rye. In addition, cheap rum was now available in the U.S. Towards the end of the nineteenth century these spirits started to replace cognac in the American lifestyle, not only in cocktails.

About the same time—in 1872 , to be specific—a plague hit the vineyards of Cognac. It happened just as a new tax on wine and spirits was put in place to fund mandated remuneration after Napoleon III lost the war with Prussia; with the addition of this tax, French domestic consumption of cognac declined drastically.

The ailment in the Cognac vineyards was a slow-growing pestilence; it took a good twenty years to destroy the vineyards. The name of the killer was phylloxera, the grapevine-destroying louse that spread throughout Europe and decimated vineyards from the late nineteenth century into the early twentieth. No grapes in France were safe, and there was no cure, no way to destroy this ravenous pest. In Cognac, phylloxera spread slowly, over the course of several decades, eventually slashing the vineyards to a fraction of their size by the early 1890s.

True cognac is unique in the world because of its grapes and the region's climate, and in the experience of the people who distil and age it—and promote it, too. But to add to their misery, Cognac's producers found that demand for their high-quality spirits had inspired people in a variety of countries around the world to produce their own brandies. Many produced inferior spirits and labelled them 'cognac'. For example, producers in both Armenia and Georgia imported stills and knowledge from Cognac in the 1870s in order to produce high-quality 'cognac' for their domestic market. In the 1880s producers in Italy and Greece began marketing their domestic brandies widely. They used different grapes, and their products may (or may not) have been very good quality, but they all hijacked the name 'cognac' to sell their spirits. So, in addition to having to figure out how to reconstruct their vineyards, the people of Cognac had to learn how to compete with other brandies in the world market. And they had to begin the difficult process of challenging foreign producers in order to defend the uniqueness of the name and identity of cognac across the globe.

Back in Cognac, without an effective, long-term treatment for phylloxera, there was only one thing to do in the late 1800s: begin wholesale replacement of the vineyards with phylloxera-resistant vines. This is the 'cure' that was finally discovered, after everything from religious to chemical to water therapies (flooding the vineyards) had been tried.

As in the rest of France, Cognac planted rootstocks imported from America, and grafted European grapes on to the roots. This worked well in Bordeaux and Burgundy with grapes like Merlot, Cabernet Sauvignon and Pinot Noir. At first Cognac's growers grafted the same white grapes they had been using to make cognac for centuries: Folle Blanche and Colombard. But they found that Folle Blanche did not do as well on American rootstock, so they began to graft more and more vines with another white grape from the area, Ugni Blanc (also known as Trebbiano). Ugni Blanc made a slightly different, perhaps more undistinguished wine than Folle Blanche, but it distilled into a wonderful cognac—either with or without the addition of Colombard. Finally, Cognac would be able to get on with its grape growing and cognac production.

Chapter 4
Armagnac and its Noble History

It seems fitting that today, the Armagnac region of central, southwest France has a timeless feel, the bucolic landscape looking much the same as it did hundreds of years ago. Armagnac is the most ancient brandy-producing area—older than Cognac by hundreds of years. The distilled spirit known as armagnac celebrated its 700th birthday not too long ago, in 2010. Why did distilling develop in Armagnac so early? And why didn't armagnac become more well known than cognac? The reasons for this are rooted in the region's geography.

Most armagnac houses have their production headquarters set in the countryside, with its rolling hills, small farms and pocket-sized pastures. They might be located near a village or out in the midst of their vineyards. A few producers are located in town centres, such as the Ryst Dupeyron armagnac cellars, which have been headquartered in the historic town of Condom for over 100 years. But what happened between the creation of armagnac in the 1300s and now? And how was armagnac initially created?

As described in chapter One, during the early Middle

Ages the technique of distilling spread from the Middle East through the Iberian peninsula into the south of France. At that time, the Western world was in the process of a slow shift from the reign of magic and mystery to the ascendancy of science. Medicine straddled both fields, and doctors prescribed the wine distillate known as brandy, medicinally.

The first person to codify the health attributes of the distillations of armagnac in print was a Franciscan friar named Vital du Four, who lived from 1260 to 1327. Du Four wrote a medical book in 1310 relating the importance of Armagnac's distilled spirit to physical health—and this was the beginning of armagnac's fame. He called this spirit '*Aguay Ardente*' (*aqua ardente*) and it is considered the direct ancestor of the armagnac of today.

Du Four's work was so important that it was kept alive in manuscript form for centuries. With the invention of the printing press, it reached a still-wider audience. Vatican archives now house a copy of Du Four's book published in 1531. In his writings Du Four lists 42 benefits of consuming a drink made with Armagnac's *aqua ardente*. The miraculous properties include curing wounds and sores; restoring memory; healing paralysed limbs; and imparting courage to the faint of heart.

Armagnac's production grew over the following centuries, keeping up with increasing demand in this part of France's Gascony region, whose landlocked geography played a major part in armagnac's development. In fact, the lack of commercially navigable rivers kept Armagnac relatively isolated from the outside world. So, though armagnac may claim to be the first French brandy of quality, until comparatively recently it was so difficult to export this spirit that it never became as famous as cognac worldwide.

Armagnac now differs from cognac in another significant way: it is manufactured in a single-distillation process. Though armagnac was initially produced in both single- and double-distilled pot stills, for the past two centuries it has consistently been made in the 'Alembic Armagnaçaise', a single squat, cylindrical still patented in 1818. This produces a fine brandy with an entirely different flavour profile from that of cognac.

This artisanal type of single distillation has another great advantage: it can be done in a still so small that it is portable. And this is what occurred in Armagnac: stills on wheels were moved from one small grape-grower to the next, allowing even ordinary farmers to make their own armagnac on their own land. The custom persists today and this, along with the slow growth of armagnac production in the past, has allowed many modestly sized producers to flourish among their vineyards. Artisanal producers on country lanes now add immensely to the charm of the region.

However, the grapes that armagnac is made from today are not the same as those that were used there until the late nineteenth century. As in the rest of France, Armagnac's vineyards suffered greatly during the phylloxera epidemic. Before phylloxera came to attack the roots and destroy all the *Vitis vinifera* (wine grape) plants, the highly acidic white grape Folle Blanche was the basis for armagnac, as it was for cognac. When the vineyard plague hit, a French breeder named François Baco developed a resistant grape that was a cross of the traditional Armagnac and Cognac grape Folle Blanche and an American grape called Noah (which had phylloxera-resistant roots). Now many producers find that this grape, with the unromantic name Baco 22A, makes the finest armagnacs—especially when planted on the sandy, gravelly soil of the best Bas-Armagnac areas.

Armagnac contains three regions, defined for the purposes of grape growing as well as production. Running basically west to east, Bas-Armagnac is considered the finest, then Armagnac-Ténarèze, then Haut-Armagnac. Today, Baco, Colombard, Folle Blanche and Ugni Blanc are the main grapes used in Armagnac. A total of ten grape varieties are allowed; the other six are the much less well known Clairette, Graisse, Jurançon Blanc, Mauzac (Blanc and Rosé) and Meslier-Saint-François.

As a spirit, armagnac tends to be fruitier and more floral than cognac. It also takes more time for armagnac's internal elements to meld together and soften with age, which means that even the entry-level tier of fine, smooth armagnac is often older than first-tier cognac, and consequently more expensive.

Though sometimes armagnac is aged in Limousin oak barrels from the forest near Cognac, when armagnac is put into barrels from its native Gascon forests, it gains a different flavour profile and an increasingly golden tint throughout its life. Armagnac today has a tiered system of ageing, with rigorously enforced regulations. However, the Bureau National Interprofessionnel de l'Armagnac (BNIA, established in 1941) is in the process of simplifying the descriptive terms for armagnac labelling. It advocates the following standards: the entry level, VS or 3-star, is defined as having had over one year of ageing; mid-range VSOP must be over four years old; Hors d'âge denotes a true, aged armagnac at least ten years old, with the age also given on the label (for example ten, fifteen, twenty-five years); and Vintage armagnacs must be at least ten years old and must list the year of harvest on their labels. The Bureau also wants to see categories which are used to grade cognacs, such as XO, Vieux and Napoléon, being phased out in Armagnac.

After bottling, armagnac is ready to drink; it does not improve in the bottle. But once opened, it keeps for many weeks, even months, when stored in a cool place. Currently, up to 500 producers and 300 cooperatives produce a total of about 6 million bottles of armagnac a year. Major labels on the shelves in the U.S. and UK include Baron de Sigognac, Castarède, Darroze, Dartigalongue, Delord, Gélas, Janneau, Laubade, Larressingle, Marquis de Montesquiou, Pellehaut, Ryst Dupeyron, Samalens and Tariquet.

While it remains a relatively rare spirit in the U.S. (a bit less rare in the UK), armagnac's popularity is on the ascendancy in China. With modern transportation, the spirit is now unhindered by the geographical constraints of trade that limited its distribution for centuries. As a prestige spirit in East Asia, armagnac satisfies a desire for full flavour in a drink and carries the status of a long history, as well as being presented in appealingly sophisticated packaging. Armagnac has taken off meteorically in China in the course of only a few years; by 2012 China's consumption of armagnac exceeded that of the U.S.

Chapter 5
Illustrious Brandies of Europe and the Caucasus

As the illustrious reputation of cognac spread during the nineteenth century, it inspired brandy production in other countries in Europe and further east, notably Armenia and Georgia in the Caucasus Mountains.

In the late nineteenth century, distillers could trade on the fame of cognac, making brandy that was less expensive to produce and that sold for much less than imported cognac. In many areas, the local brandies were even called 'cognac', a term that has been slowly transitioning to 'brandy' around the world after much hard campaigning by cognac producers. Outside Cognac, most brandy manufacturers made their spirits with local grapes, but some imported stills, techniques and even grapes from Cognac.

In Germany, for example, where cognac carried a significant reputation, many brandy producers imported some or all of their brandy grapes from Cognac. Even today, top German brandy producers Asbach Uralt and Dujardin bring in grapes from the Charente (Cognac) region—as do some brandy producers as far away as Asia, including Russia and India.

In Italy, brandy has been distilled since the sixteenth century, but the country does not have a designated geographic region for brandy production. One of Italy's most distinguished brandy companies has been around since 1820 when a Cognac native called Jean Bouton found his way to Emilia Romagna and discovered Cognac's Ugni Blanc grapes (known as Trebbiano) were grown there as well. He became known as 'Giovanni Buton' when he established his distillery and began producing Vecchia Romagna brandy—which remains one of the top-selling Italian brands today. Stock, an Italian brandy producer established in 1884, makes some of the best-known brandies in Europe. Though its heyday was in the 1960s and 1970s, it is still extremely popular in Italy and other countries.

Another top Italian commercial brandy is made by Fratelli Branca, whose Stravecchio Branca brandy is a household name; a common Italian recipe for a sore throat treatment is a dollop of this brandy in a glass of hot milk. The company began making brandy in 1888, with a unique process: before bottling, Fratelli Branca blends in up to two-thirds of an existing brandy which is kept ageing in a 'mother barrel', so their finished blends contain brandies aged for anywhere from three to ten years. Fratelli Branca also produces a premium distilled brandy called Magnamater. Other Italian companies produce high-end products aimed at connoisseurs, as well: wine- and grappa-producing companies that also make brandies include Villa Zarri, Marchesi de'

Bianchi and Giori, and the winemaker Marchese Antinori, the sparkling winery Bellavista and grappa producer Jacopo Poli.

Moving across the Mediterranean to Greece, we find an interesting aside in the most famous Greek 'brandy', Metaxa. (This is a drink many travellers have intense memories of, from youthful Mediterranean holidays.) Spyros Metaxa began producing his eponymous spirits in 1888, and Metaxa's fame has spread around the world. His spirits were a welcome departure from the harsh, local product of their time. However, he combined his distillate with sweet muscat wine and herbs, so strictly speaking, classic Metaxa is not a (classic) brandy.

Further east, quality brandies are also produced in the Caucasus, in Armenia and Georgia. Long hidden from Western view, first by geographical remoteness and then by the Iron Curtain, these countries have prolific brandy companies with long traditions of quality brandies made with Charentais (Cognac-style) distillation and ageing, which they supplied to the tsars of Russia. Legend has it that when the Bolsheviks breached the Winter Palace in 1917, the entire revolution paused for a week while the revolutionaries drank up the Tsar's incredible brandies.

Until a couple of decades ago, Armenia was the main designated producer of brandies supplied to Russia and other countries of the Soviet Union. When the USSR was disbanded, distribution networks disappeared overnight and the market for Armenian brandy collapsed. Currently, the

Armenian industry is rebuilding itself; its former markets have stabilized and its producers are looking towards new markets.

Today there are three large brandy-producing companies in Armenia, and some struggling with privatization. A few are still located in the capital city of Yerevan, generally with updated facilities. Others are located on the surrounding, high, arid plateaux of this mountainous country. The major companies are Ararat, Noy and Proshyan, with other producers such as the formerly-great Vedi-Alco at the start of a comeback.

Confusingly, both Ararat and Noy are sometimes referred to as the Yerevan Brandy Company. Ararat is named for the biblical mountain where Noah landed; Noy is the Armenian version of Noah. Both companies claim to have launched the brandy industry in Armenia in 1877. Originally, a brandy distillery and ageing cellars were built on the site of a sixteenth-century Persian castle that commands a view of the approach to Yerevan. In 1899 this company was acquired by a Russian industrial magnate and brandy promoter named Nikolay Shustov, who is considered the father of the vital Armenian brandy industry. This industry played such a significant part in the country's history that a stamp with Shustov's picture was issued as recently as 2007.

Shustov's company became the official supplier to the royal court of Tsar Nicolas II in 1912. Throughout the Soviet era, these brandies were in great demand in Russia and other Soviet states. Stalin is said to have introduced Winston

Churchill to Armenian brandy during the talks at Yalta in 1945. And Churchill apparently enjoyed it so much that it is said Stalin sent him a case every year for the rest of his life. (Another Churchill story follows, in the section on Georgian brandy.)

During the Soviet era the name of Shustov's company was changed to Yerevan, with production headquarters and ageing cellars moved to a new, updated facility in the city in 1950. In 1998 the company was acquired by the international wine and spirits conglomerate Pernod Ricard and the company is now known as Ararat. Its modern production facility uses cognac-style distillation for its famous brandies. The Ararat company is now supplied by 5,000 grape growers and produces 5.5 million bottles a year, 92 per cent of which are exported to Russian and Baltic nations.

The Ararat cellars also hold a 'Peace Barrel', which melds 1994-vintage spirits from Armenia and Azerbaijan to represent the year of the ceasefire over the contested Nagorno-Karabakh region. The barrel was dedicated in 2001, and waits to be opened when there is a formal treaty of peace in the region.

After the demise of the Soviet Union, private investors scraped together enough money to start up production again at the original Shustov facility in 2002 They named their new venture Noy, and the company logo shows Noah's ark with the date 1877 on it. Apparently, the former company never moved all their original stores to the new location because today, deep underground, brandies that are close to 100 years old can be found in Noy's cellars. The Armenian brandy industry has come full circle: in 2011 Noy created a new line of brandy to supply the Kremlin.

Proshyan, according to the owners of the company, is an ancient, noble name; it is also the name of the village outside Yerevan where this company is headquartered. There was a 'Proshyan Brandy Company' founded in 1887, but the current company is not a direct descendant. It was not in existence in the Soviet era, and though some of the company's production equipment dates from that time, it is now being augmented by new machines from Italy. A new glass and marble office building opened in 2012 to crown the old production yard. A thoroughly modern corporation in feel, Proshyan produces branded labels for 500 European supermarkets and sells to Russia, Germany, the Baltic nations, South Korea and China. Yet one of their claims to fame are their traditional, ornate bottles in fascinating shapes from roses and ships to swords and dragons. These 'souvenir' bottles account for a good 20 per cent of the market for Proshyan.

In direct contrast to this thriving enterprise is Vedi-Alco, located at an old Soviet plant out in the countryside a couple hours drive from Yerevan. It was taken over in 1996 by a group of workers desperate to improve their fortunes after the fall of the Soviet Union. The original factory was established in 1956, and the facilities still have a mid-twentieth-century Soviet feel. In 2000 the company started to distil brandy there again, and it also acquired some aged spirits. Currently the company does batch distillation in a column still, but hopes to be able to add a double-distilling system like the one it used to have—though repairing the roof has taken priority most recently. Demand is just starting to pick up, and Vedi-Alco brandy is now exported, mainly to the Russian market.

Culturally, brandy is considered a dessert beverage in Armenia. It can be served with chocolate, with oranges and apples, or with ripe, fresh peaches in season. Traditionally—as in many areas of the West—cigars also accompany brandy after dinner. Etiquette dictates that, when offered brandy in a snifter, a guest should be able to lay his glass down on its side and nothing will spill out; this signifies that the host has poured the correct amount for sipping, around 50 ml (about 1.7 fl. oz).

Currently, thirteen (mainly white) grapes can be used to make the wine for Armenian brandy: Azeteni, Banants, Chilar, Garan Dmak, Kakhet, Kangun, Lalvari, Masis, Meghrabujr, Mskhali, Rkatsiteli, Van and Voskehat.

During the Soviet era, the adjoining country of Georgia was known for its wine—though brandy has also been produced here since the late nineteenth century. For the past 130 years, Sarajishvili has been the most important Georgian brandy, founded by a Georgian native, David Sarajishvili, in 1884. After studying chemistry in Germany and distilling in Cognac, he had to return home after the death of his father. He sought out grapes that had similar characteristics to those used in Cognac from among the 500 varieties native to

Georgia, selecting them from various regions of the country: Chinuri, Goruli Mtsvane, Kakhuri Mtsvane, Rkatsiteli, Tsitska and Tsolikouri. It has become customary for Georgian brandy producers to source grapes from around the country.

Sarajishvili also imported two additional significant elements from France: a copper alembic still from Cognac, and a connection to Camus Cognac, the area's oldest family-owned production company. This connection endured for decades but was severed during the Soviet era and then re-established afterwards, with the father of the current head of Camus.

During the Soviet era the name of the firm was changed to the Tbilisi Brandy Company, and its products were often requisitioned by the authorities. Despite this, the distiller was somehow able to sequester a few historic barrels, including several made by David Sarajishvili himself in 1893 and 1905, which are in the cellars today. A little of the contents of these historic barrels is now blended into the company's top-of-the-line brandies.

Headquartered on a garden-like property established in Tbilisi in 1954, the Sarajishvili company was privatized in 1994 and continues to produce a range of aged brandies, double-distilled in its copper stills. Its chief technologist David Abzianidze carries the legacy of Sarajishvili brandy as a historian of the company and a forward-thinker for its products. However, he is not interested in having his brandies used in cocktails—as popular as this may be in other countries.

Abzianidze tells a (rather familiar) story learned from

his predecessor: when Stalin gave Winston Churchill some Sarajishvili brandy during the Yalta talks in 1945, Churchill thought it was as fine as cognac, choosing it as the best brandy there.

Today several other Georgian companies are producing brandies, capitalizing on the status of Sarajishvili in particular and Georgian brandy in general. They may use grapes from any part of Georgia, and they bottle and market their products to countries familiar with Georgia's reputation. The wine company Tiflisi Marani is one—though it is better known for its wines at present. Another is KTW (Kakhetian Traditional Winemaking), a young company—literally—staffed by a group of fresh, energetic people. It was founded in 2001 and already has had great success in producing mid-priced wines and brandies for Eastern European and Baltic countries. KTW has created a niche for itself with old-fashioned-style packaging, and many of their brandies have a homemade look in both bottle and flask shapes.

Though most of these Georgian and Armenian brandies have not yet reached the U.S. or UK markets, it is probably only a matter of time until more of them appear on the shelves—but perhaps they will reach the eager Asian market first.

Chapter 6
Great Spanish and Latin Brandies

Though brandy is often considered to be a French drink, Spanish and Latin American cultures have strong brandy traditions that date back many centuries. In this chapter we take a look at brandies and their heritage on the Iberian Peninsula—and at the very important contribution of Peru.

Spain is the origin of a noble, Old World brandy that is not as well known as it should be today: brandy de Jerez. It is produced in the Jerez region, where sherry also originates. Jerez is in southwestern Spain, not far from the southern Atlantic coast near the Straits of Gibraltar. The sensory distinctiveness of brandy de Jerez comes not only from its geographic location but also from the unique ageing system for the brandies produced there. Brandy de Jerez and sherry are both aged with the same method, the *solera* system (discussed below). This practice imparts a unique flavour and aroma profile to the brandy: hints of marine salinity, vanilla and toasted caramel, carob and coffee notes, underlying layers reminiscent of mature oak and yeast.

Wine and other products have been traded in this part

of Spain since the Phoenicians roamed its seas from around 700 to 500 BC. Wine was also exported during and after the time of the Roman Empire, until the Moorish occupation of the Iberian Peninsula from AD 711 to 1492. Though the Moors traditionally did not drink wine, they made use of the grapes in the existing vineyards to produce an alcohol distillate for use in medicines, cosmetics and perfumes.

When the Moors were driven back towards the south—and eventually out of Iberia—their distilling technology remained. Local medical and alchemical specialists employed the process to create their own 'waters of life' and 'waters of the spirit'. In fact, *aguardiente* (from *aqua ardens*, the 'burning water' of the Middle Ages) is still the generic Spanish term for brandy.

Though the first reference to brandy in Jerez appears in 1580 in relation to a tax on spirits, brandy could have been manufactured there earlier. In this Spanish region, as in France, it was the Dutch who exploited the manufacture of brandy beginning in the late sixteenth century. Today the clear spirit that comes off the pot stills to be aged into brandy is still called *holandas*. And many streets, pavements and patios in Jerez de la Frontera are lined with the small, rounded stones brought from Holland centuries ago as ballast for ships that took on cargoes of distilled spirits in Jerez.

Through the eighteenth century, this Spanish brandy was shipped unaged; in fact, under the *gremio* (grape growers' league) law, *holandas* was required to be shipped out every year so the growers and distillers could empty their cellars before the next harvest, and get paid for their products promptly.

Legend has it that in the early nineteenth century, a shipment of *aguardiente* was put into used sherry casks, ready to be shipped out. But the ship left without these casks, which were only discovered later. After tasting this spirit, the producers realized that it had improved in the barrels, and this is how barrel-ageing brandy de Jerez began. The producer was the famous company now known as Pedro Domecq, and the year was 1818.

Production and trade of brandy from the Jerez area increased steadily through the nineteenth century. The brandy was originally made with the region's native Palomino grape (also used in sherry), but with the brandy boom in the late nineteenth century, producers had to look farther afield. They found the required characteristics in Airén grapes, grown in the centre of Spain in the Castilla-La Mancha region.

A large proportion of the grapes for brandy de Jerez is still grown in Castilla-La Mancha. Wine from these grapes is made in the La Mancha city of Tomelloso, where many brandy producers also own distilleries. However, brandy blending and ageing is always done in the Jerez area, specifically in the designated coastal 'sherry triangle' area bounded by the towns of Jerez, El Puerto de Santa Maria and Sanlúcar de Barrameda. Some brandy is also distilled in Jerez.

For brandy de Jerez, both pot stills and column stills are used. Technically, when a still is heated by steam it is called an alembic, and when it is heated by wood it is an *alquitara*. The spirits from both types of stills are classified as *holandas de vino*.

The Jerez barrel-ageing warehouses (bodegas) and their offices are generally whitewashed adobe structures a few storeys high, with orange-red terracotta roof tiles. Many are set up like miniature versions of a nineteenth-century villa located in the heart of the city of Jerez: warrens of buildings behind walls, all charmingly accented with deep-red bougainvillea. Airy and pleasant year-round, each hive of business contains white buildings connected by brown cobbled lanes, often interspersed with gardens or open squares.

Brandy de Jerez barrels (previously used for sherry) are made of American oak, a custom that began when the region first became a prosperous Atlantic trading zone. Barrels that have been used to age different styles of sherry—from sweet to dry—are selected to shape the aromas and flavours of the brandies.

Brandy de Jerez is produced solely by the *solera* method, which according to legend was due to another fortunate accident. In 1870, numerous barrels of brandy were left unclaimed in the corner of a bodega. Someone discovered the barrels in 1874, but they could not sell the entire quantity of such old brandy right away, so they began mixing it, filtering down some newer brandy into the barrels to fill the space left by normal evaporation. This was so successful in contributing complexity and finesse to the aged spirit that the *solera* method has been used ever since.

In the *solera* method, barrels of brandy are stacked according to age in the high-ceilinged bodegas. The top row of casks contains the youngest brandies, and each successive row of casks contains increasingly older brandies down to the floor level (*solera* level). When some of the matured brandy is removed from the floor-level casks for bottling, the floor-level casks are topped up from the next row above, and those from the row above of them, and so on. The rows above the *solera* are called *criaderas*—a charming word meaning 'nurseries'. Commonly, there are at least three or four vertical rows of casks ageing in the bodegas, with plenty of air space above.

During the dry heat of summer, the red-brown clay floors of the bodegas are sprinkled with water to maintain humidity and limit evaporation from the barrels. Since the percentage of brandy evaporation from barrels can reach up to 7 per cent per year, at some bodegas systems for temperature and humidity control are starting to be introduced. Though these may help a company's bottom line, it remains to be seen whether the brandies will retain their unique, regional aromas and flavours without the natural breezes blowing through the bodegas.

In Jerez, the top sherry producers also tend to produce the top-quality brandies. However, they do not make the bestselling brandy. That distinction belongs to Bodegas Terry with its Centenario brandy. Terry was founded by an Irish family in the mid-nineteenth century. They named their brandy 'Centenario' when they established new facilities in the nearby town of Santa Maria at the beginning of the century—the twentieth century, that is.

Another of the most famous brandies from Jerez is the ubiquitous Cardenal Mendoza, produced by Sánchez Romate since 1887. Originally made for the owners' private consumption, it was soon commercialized and is now known around the world.

Some bodegas buy spirit to age, while others have the facilities to do more of their own distilling, such as the 150-year-old González Byass and the eighteenth-century Pedro Domecq (now part of Bodegas Fundador). Occasionally, a newer company appears, like Bodegas Tradición (established in 1998), which has been buying stores of brandies from a variety of sources to further age, blend and bottle under its own label.

Though brandy has been made in Jerez for centuries, the Consejo Regulador (Regulatory Council) for brandy de Jerez was established only in 1987. According to the Council, three styles of brandy de Jerez may be defined, by their increasing amounts of 'volatile components' and also by the classification of age: brandy de Jerez Solera, aged for an average of a year and a half; brandy de Jerez Solera Reserva, which is aged for three years on average; and brandy de Jerez Solera Gran Reserva, which averages ten years of age.

Small amounts of brandy have also been made in other areas of Spain since the boom in the late 1800s. In the north-eastern region of Catalonia, brandy was historically one source of the calories people in this mountainous terrain needed to survive in the harsh climate. As late as the end of the twentieth century—and possibly still today—a custom existed whereby workmen would stop in a café in the morning on the way to their jobs to have a coffee topped up with brandy, for energy.

An example of a Catalonian brandy producer is Mascaró, a company founded at the end of the Second World War. Historically, many Catalonians became rum distillers in the Spanish colonies, and exported their products back to Spain. But during the Spanish Civil War and in subsequent disruptions of trade through the Second World War, Spain stopped importing spirits. So Catalonians had to start distilling their own spirits at home with the materials on hand, namely wine.

Located near the French border, Catalonians such as Narciso Mascaró (the son of a wine merchant and distiller) chose to use the Charentais or Cognac method of double distilling. The Spanish sparkling wine cava is also made in this area, and cava grapes lend themselves readily to distilling. The Parellada grape, which has high acidity and delicate aromas, is the basis for many Catalonian brandies; Macabeu and Xarel·lo are also used.

Portugal, also part of the Iberian Peninsula, produces some brandy too, notably in Lourinhã. This historic area of brandy production received DOC (Denominaçao de Origem Controlada, or PDO, Protected Designation of Origin) status only twenty years ago, though there has been a tradition of brandy distilling there since the eighteenth century. Lourinhã is located in a wine-making region north of Lisbon and the brandy made there is referred to by the traditional Portuguese term, *aguardente*.

Across the Atlantic, one of Spain's former colonies, Mexico, has followed the lead of its erstwhile mother country in brandy production. Mexican brandies are also made by the *solera* method, though in general they have not been considered as fine as Spain's brandies. Until the twenty-first century, almost all of Mexico's wine grapes were used for brandy distillation. Its most famous brandy is Presidente, which is produced in Mexico by Pedro Domecq of Jerez and also exported to many countries around the world.

Until the beginning of the nineteenth century the Philippines were governed by a colonial administration headquartered not in Spain but in Mexico, and Mexico has therefore been an influence on Filipino culture. Both Spanish and Mexican brandies have made their way to the Philippines, where Spain's Fundador is also a very important brandy. Local brands Emperador and Generoso are the other top sellers there.

Meanwhile, back in the sixteenth century, there was a parallel development of brandy going on in South America, especially in Peru, home of the brandy called pisco. Though many people do not realize it, pisco is a brandy. Pisco differs from other brandies in that it is a white (clear) spirit, and no water is added after distilling. In other words, the spirit must come off the still with the desired level of alcohol for bottling, which is between 38 and 43 per cent. Brandy de Jerez, cognac and most other brandies are distilled to a higher alcohol level before water is gradually blended in during the months before bottling.

Pisco originated as early as the beginning of the seventeenth century along the southern coast of the country, where grapes had been grown since the mid-sixteenth century. It is named after the port city of Pisco, the shipping point for much of this brandy. Spanish explorers and settlers had brought grapes from the Canary Islands and from Spain to plant near Peru's south coast because wine was part of their religion and culture. This area proved so prolific for grapes that in a short time wine was being exported from Peru to Spain. But Spanish wine-makers strongly protested against competition from colonial imports so a law was enacted in 1641 banning the importation of Peruvian wine. So Peruvian winemakers began distilling their wine into brandy (*aguardiente de vino*) and exporting that to Spain, where it found a ready market.

Originally, Peruvian brandy could be distilled in either of two types of stills. The *falca* looked like one of the early Moorish or medieval stills: a simple container for heating and a long pipe for condensing the vapours. Peru's other type of still, its alembic, was more sophisticated, like the European alembic stills with their long, spiral necks of copper tubing. In this latter process heads and tails were discarded and the pure heart of the pisco vapour remained, to be cooled and poured into clay *botijas* for shipping.

Though pisco was imported into the San Francisco area during the seventeenth and eighteenth centuries, a demand for pisco intensified when the Gold Rush pioneers arrived in California in the mid-nineteenth century. Pisco's renown only increased during the course of the rest of the century. In the early 1900s, San Francisco was the seat of a pisco craze: people couldn't get enough of it. Often they drank it in a Pisco Punch, said to have been invented in San Francisco. Pisco Punch, a concoction with lemon, sugar and pineapple, also spread across California to Nevada, according to newspaper articles of the time.

As wildly popular as pisco and Pisco Punch were in the West in the early part of the twentieth century, Prohibition effectively destroyed the pisco market in the U.S. Afterwards, pisco declined into a notoriously rough, cheap spirit—the type of drink one would expect at a bar in an old western movie set in California.

But pisco was not dead; in fact, at the turn of the millennium it was ready for resurrection. Authentic pisco was being rediscovered by travellers to Peru, and producers again began to have a market for a finer version of pisco. Though pisco had been made for centuries, it was only in 1999 that Peruvian producers created their own Denomination of Origin rules for the area of production, the types and quality of grapes, and the distillation and ageing of the spirit.

Barely a decade into the twenty-first century, pisco began a new upswing in popularity, especially within the current mixology movement. Today the spirit is produced as pisco puro (from a single grape variety) or as pisco acholado, from more than one grape, and usually with grapes that are both aromatic and non-aromatic.

Pisco can also be made either from wine that is fully fermented, or from wine that is partially fermented (still somewhat sweet); and sometimes it is made from a mixture of the two. Pisco can be made from eight grapes: the aromatic grape varieties blended into pisco are Italia, Torontel, Moscatel, and Albilla, while non-aromatic piscos (which are more typical) tend to be made from the grapes Quebranta, Mollar, Negra Criolla, and Uvina. Quebranta is by far the dominant grape in Peru's pisco.

As much as Peru has been identified with pisco for centuries, today it is not the only country that makes this spirit: Chile has launched an effort to compete on the world pisco market. Historically, a pisco-style brandy had been made in Chile, though even as recently as a decade ago in Chile it was common to look to Peru for the better piscos. But not any more.

Chilean piscos, however, are quite different from Peru's in aromas and flavours—Chile's tend to be softer and more fragrant. Currently, several producers in Chile are making very good piscos in cognac-style double-distillation stills with aromatic Muscat grapes, notably companies like Kappa (from the winemaking Marnier-Lapostolle family) and ABA. They are using high-quality wine grapes, and it shows in the final product.

Though the pisco industry is still dominated by Peru,

Chile actually has had its own pisco denomination since 1936, when a town in the grape-growing region of Elqui changed its name to 'Pisco Elqui'. And Chilean producers are also experimenting with another cognac element: barrel-ageing.

Fine brandy-making is found somewhat randomly in the rest of Latin America. In Argentina, for instance, Ramefort Coñac was started with advice from cognac producers, and its brandies are made in the cognac style. On the other hand, Bolivia has its own special style of brandy called *singani*; distilled mainly from high-altitude-grown Muscat grapes, it is a unique spirit.

After this exhaustive tour of Spanish-inspired brandies, it is a toss-up whether one would want to relax with a sipping-glass of brandy de Jerez, or a Pisco Sour cocktail—before we continue on our world tour of brandy.

Chapter 7
Australia and South Africa

Whether they brought the taste for it from their home country or developed it abroad, inhabitants of the British Empire were robust consumers of brandy. And, as mentioned earlier, brandy was also regarded as a household staple for medicinal purposes—a state of affairs that lasted well into the twentieth century.

Located far from Europe, British Commonwealth countries, most notably South Africa and Australia, made their own brandies for domestic consumption. In Australia, Château Tanunda is an example of a well-established brand that has been recognized for more than a century. This brandy comes from one of the first areas in the country where grapes were planted, the Barossa Valley in South Australia. Château Tanunda was established there some decades later, in 1890, but the grandeur of its estate and the company's savvy salesmanship ensured that Tanunda's brandy became known as 'the Commonwealth's hospital brandy'.

Brandy was firmly believed to be a cure for all types of problems. In fact, brandy is credited for a sports career, when in 1896 the famed cricket batsman Frank Iredale was 'Saved from Failure by Brandy and Soda', as a *Perth Daily News* headline declared.

The grapes planted in the Barossa Valley in Australia in the mid-nineteenth century were used for both wine and brandy. Syrupy-sweet fortified wines, sherry-style wines and heavy brandies were very popular in Australia from the mid-1800s. Unfortunately, it seems the Australian perception of brandy has been trapped in the past—few younger people today are interested in brandy. Brandy is considered an accept- able spirit, but in a curiously limited way: it is consumed mainly by women over the age of 40, say Australian brandy producers.

Though the Angove family had been making wine since 1855 in Australia, they started growing grapes specifically for brandy in 1910. Carl Angove opened the first industrial distillery there in 1925. He followed the French and Spanish model, using grapes that were fairly neutral but could be harvested with good acidity: the table grape Sultana, as well as the French Colombard and Spanish Palomino. Angove brandy was a bit of a departure from the heavy, sweet style most Australians had been used to in wines as well as in spirits. It was considered more of a clean, cognac-style brandy, and it was very well received—as it is today. Angove's St Agnes brand has over 70 per cent of the market in its native state of South Australia, and 40 per cent of the country's brandy market. And Angove is reportedly the only company still making their brandies in a Charente-style pot still, double-distilled.

In addition to Angove and Château Tanunda, the most popular brandies in the country are Hardy's Black Bottle brand and the Woolworths-distributed brand from France called Napoleon 1875; French cognac by Rémy Martin also does well there. Brandy consumption in Australia has been extremely stable for some time. And there has been no traction in the movement for brandy cocktails. But very recently there has been a new development: China. Enamoured with European brandies, the Chinese are also looking at other sources to supply their high-end brandy appetite. Suddenly, they have become interested in, for instance, Angove's XO, and this may be the beginning of a new trend.

In contrast to Australia, in India the major brandy consumers are men. The high-end spirits they drink are usually imported from Cognac. But at the middle to lower range, the spirits are likely to have been manufactured in India. Lacking a significant amount of native grapes and grape wine, manufacturers in India often distil their spirits out of sugar products—so technically these spirits would all be rums, not brandies or whiskies. But they are coloured, sometimes aged and/or flavoured, labelled as brandy or whisky and sold to a public that has become used to this style of spirit. So-called 'brandies' that are manufactured in other high brandy consumption countries in Asia (such as Malaysia and the Philippines) are also more than likely to be produced by the same rum-like process domestically.

In contrast, another former Commonwealth country,

South Africa, has a strong cognac-style brandy orientation due to its roots as a Dutch colony dating back to 1652. Early settlers had planted grapes in South Africa by 1659. The popular account of brandy's beginnings in South Africa mentions a Dutch ship named *De Pijl*, which first distilled brandy there while it was anchored offshore on 19 May 1672. As mentioned earlier, the Dutch were responsible for distilling wine into brandy in several areas of France in the seventeenth century, so they had the knowledge and equipment to do this wherever they found suitable grapes.

With plenty of white grapes to distil, South Africa's brandy industry grew in parallel to its wine industry. Brandy is made here mainly from Chenin Blanc and Colombard (which in South Africa is often spelled Colombar, without the final 'd').

For several centuries, virtually all of South Africa's brandies were designated for domestic consumption. One notable early brandy enterprise was Van Ryn, established in 1845. After buying out F. C. Collison (which was established in 1833) it has laid claim to being the oldest continuous brandy making concern in the country. Van Ryn distils brandies in a cognac style, and even has its own on-site cooperage.

South Africa's largest brandy producer is KWV (Koöperatieve Wijnbouwers Vereniging van Zuid-Afrika), a company that was founded in 1918; it became a winemaking cooperative in 1923 and began brandy production in 1926.

During its history, this cooperative has been both a private and a public company at various times. As the regulator of the South African wine industry until 1977, KWV exported its bottled brandies because it was not allowed to compete on the domestic market—though it did sell brandy in bulk to other companies which then aged and bottled it. Currently, it is a private company that both exports and sells domestically—and remains very important in the South African brandy and wine market.

Another significant company, Klipdrift, started out in 1938 as a small, backyard distillery but rose in record time to become South Africa's best-known brandy. Unfortunately, this is one of the brandies that fuelled the decline of South African brandy in both quality and perception. By the middle of the twentieth century, much South African domestic brandy had become the equivalent of cheap rum in America. 'Klippies and cola', like rum and Coke in the U.S., represented a basic, low-end drink. And then there was '1-2-3' (also sometimes called '3-2-1'), the downscale version of a fun evening: 1 litre of brandy, 2 litres of cola and a 3-litre Ford.

But in the last few years, changes have begun to be felt in the industry. Just as rum brands in the U.S. have worked very hard to upgrade their image from rum-and-Coke status, South Africans have begun to prove their spirits are worthy of a better standing in the spirits industry and have introduced quite a few fine, aged brandies to both their domestic and export markets.

The South African Brandy Foundation was founded in 1984. It has standardized brandy production into four categories. The first level is called Blended Brandy and is made to be used in mixed drinks; it requires a minimum of 30 per cent pot-still brandy aged in oak for at least three years, and the remainder can be neutral, unaged spirit. The second level, Vintage Brandy, requires at least 30 per cent pot-still brandy with up to 60 per cent column-still spirit matured for at least eight years, and up to 10 per cent wine spirits (unmatured). The third level, Potstill Brandy, must have a minimum of 90 per cent pot-still brandy and a maximum of 10 per cent neutral unaged spirit. The fourth category, Estate Brandy, must be entirely produced, aged and bottled on one estate; it is always labelled with the word 'estate' as well as the type of brandy.

Label terms like VS and VSOP may be used, but they do not have the same ageing designation as in Cognac. South African brandies must be aged in 340-litre French oak casks for a minimum of three years. *Solera* ageing is also allowed in South Africa. Though there is not a designated brandy production area, grapes for the South African brandies tend to come from major wine-grape growing areas including Worcester, Olifants River, Orange River, Breede River and Klein Karoo.

Brandy continues to be the top-selling spirit in South Africa. In 2008 a new annual festival called Fine Brandy Fusion was launched, mainly to attract attention from younger consumers and to position brandy as high-end and glamorous. Interestingly, today the South African Brandy Foundation website also harks back to the origins of distilled spirits with this lyrical description: 'Making brandy is akin to alchemy, when nature's elements—earth, wind, water and fire—are transformed into gold.'

Chapter 8
Brandy Made in America

Historically, brandy was considered a household necessity in the United States, for uses ranging from drinks to medical remedies; in fact brandy was classified as a medicine until the early twentieth century. In the eastern U.S., cognac and other French brandies were imported. But in the west of the country there was little commercially available brandy, except for Peruvian pisco.

After the population explosion in the west in the late nineteenth century, there was a new hunger for brandies there. Two of the legacy brandy companies in the United States began their production in the 1880s: Christian Brothers in 1882 and Korbel in 1889. Francis Korbel was a Bohemian immigrant who, with his brothers, was drawn to California's unlimited opportunities after the Gold Rush. He found land north of San Francisco, where he began making sparkling wine. After mastering wine production and marketing, Korbel went on to do the same with brandy.

Korbel may—or may not—have been influenced by the success of a lay religious order that had begun crafting and selling brandy a few years earlier, to finance their educational mission. Production and distillation by Christian Brothers (and many later entrants into California's commercial brandy business) were headquartered in the state's fertile inland valley, where a large percentage of California's fruits and vegetables were grown. Here, start-up brandy producers had access to quantities of inexpensive grapes and many of them began production with common table grapes like Thompson's Seedless and Flame Tokay. Later, more wine grapes were raised in this region, some of which were also used in brandy production.

Brandy producers typically tried to copy cognac's selection of raw material: producers believed they had to use grapes that did not have the potential to make great wines in order to produce wonderful brandies. That was true to an extent, especially if the grapes had aromatic components like florals that could transcend the distillation process. And these types of grapes could be picked earlier, with higher acidity and lower sugars, so they could be bought more cheaply: less time on the vine meant less time and expense caring for the vineyards before harvest each season.

Several decades after commercial brandy production began in California, Prohibition put a stop to winemaking. Because of its supposed medicinal qualities, cognac was the only liquor allowed to be imported into the U.S. during Prohibition. Brandy had ceased to be classed a medicine in the U.S. a few years before Prohibition, but many doctors continued to prescribe it and brandy was considered as much a part of a household medicine kit as Band-Aids are today. There are no comprehensive records of brandy being produced in the U.S. during Prohibition, but home brandy-making kits were available then—though people used whatever fruits they had available for fermentation.

Just as Prohibition nearly killed the consumer wine industry in California, it also left brandy production in need of its own restorative after Repeal. A number of individuals from across California stepped into the breach and began distilling and distributing commercial brandies again. These producers included Giovanni Vai in the Cucamonga Valley (1933), Antonio Perelli-Minetti in Delano (1936), George Zaninovich in Fresno (1937), and two brothers, Ernest and Julio Gallo (1939). Apparently, the E. & J. Gallo company got into the brandy business because there were a few bumper crops of wine grapes in the years right after Repeal. When there is more wine than can be sold, it is common to distil the wine into spirits for a variety of uses. Brandy, as an unaged spirit, was also used to fortify the sweet wines that were very popular in the U.S. for most of the twentieth century.

At the beginning, Ernest Gallo bought a few thousand barrels of brandy from a friend, to help him out. Then, in 1949, the Gallo company decided to use its 'surplus' wine to produce Gallo brandy. They produced a new product, Eden Roc brandy, in 1967. The company released Gallo brandy again in 1973, when they opened their own distillery in Fresno. Eden Roc was discontinued in 1975, when the expensive-looking E&J brandy was introduced. The company used the traditional Cognac grape Colombard, along with some Chenin Blanc, Grenache, Barbera and Moscato. E&J brandy was first distributed nationally in 1977 at that time Julio Gallo began producing his own single-variety brandies as well.

Christian Brothers and Korbel started producing brandy again after Prohibition. Additional large corporations also began making brandies to fill an increasing demand in the U.S. Many of these companies were also the producers of the most popular mid-century American wine brands such as Almaden, Italian Swiss Colony and Paul Masson. In the mid-1950s more than a dozen producers in California were making fashionable lines of grape-wine-based brandies. Large U.S. distributors also stepped in to fund and/or partner with the producers in this profitable industry; the four largest distributors were Seagram's, Schenley, National Distillers and Hiram Walker & Sons.

California brandy was in such demand for the next few decades that the biggest companies found it expedient, after distilling, to transport the brandy to Kentucky, where there were giant sheds full of used bourbon barrels that could be used to age the brandies. Though they advertised their ageing, producers generally omitted to mention where their 'California brandy'was aged.

With increased production came decreases in production quality. Though many companies started out hand-crafting brandies in cognac-style pot stills for part or all of their distilling, most moved to industrial, high-volume column stills (nothing like the small, artisanal column stills in Armagnac).

They had to keep up with a demand that was not very quality-conscious at this time. After the middle of the twentieth century a confluence of factors nearly sounded a death knell for the reputation of California brandies: lesser-quality domestic production, a generation of young people who rejected their parents' drinks, and an increase in people of all ages who travelled and experienced fine brandies in other countries.

But there remained—and still remains—a faction of U.S. consumers favouring stalwart U.S. brands like Christian Brothers, E. & J. Gallo, Korbel and Paul Masson. With their Kentucky associations, two of the 'big four' U.S. brandy producers were eventually acquired by conglomerates that had originated as producers and/or distributors of bourbon: Christian Brothers by (the appropriately named) Heaven Hill and Paul Masson by Constellation Brands. But with many of these brandies sold at low-end prices, the industry's status also declined—deservedly or not. California brandies had fallen out of favour with the young and the elite by the 1980s.

This is where the situation stood for several decades, until the comparatively recent rise in cognac's popularity. Some of the long-established U.S. brandies have risen with the tide, and all have begun to repackage and reposition their brands. For example, Gallo started producing an alembic brandy in 2003. With 43 per cent of the market, the company is currently riding the crest of a few recent trends: an increase in younger consumers, and an increase in women opting for brandy.

As will be discussed, a more positive perception of U.S. brandy may be influenced by the release of new high-end products, by an awareness of brandy that has overflowed from the burgeoning artisan spirits movement, and by a new appreciation of cognac and brandy developed within this century's mixology craze.

Chapter 9
Everything You Need to Know about Cognac

When we last immersed ourselves in Cognac, it was early in the twentieth century and the region was recovering nicely from the European grapevine plague that had decimated its vineyards. At the same time, cognac was threatened by brandy producers who had sprung up in other countries during the late nineteenth century—and who were also calling their spirits 'cognac'.

In order to characterize the uniqueness of their brandy, the people of Cognac first sought to create a delimited area for cognac vineyards. Then they codified the production and ageing requirements for their spirits. Later, they went after trade agreements with other brandy-producing regions, attempting to restrict the use of the word 'cognac'—a crusade that continues today.

Based on the most comprehensive land analysis of the region (from 1860), the first version of the delimited area was completed in 1909. Cognac became an AOC (Appellation d'Origine Contrôlée, controlled designation of origin) region in 1936, and Cognac's borders were finalized in 1938 with six growing areas (*crus*): Grande Champagne, Petite Champagne, Borderies, Fins Bois, Bons Bois, and Bois à Terroirs (also known as Bois Ordinaires). Fine Champagne is an additional appellation but not a growing area; it refers to a cognac blended from Grande and Petite Champagne grapes, with at least 50 per cent from Grande Champagne.

'Champagne' here means 'countryside' and it is a chalky soil whose vineyards provide the grapes that make the best cognacs. Grande Champagne vineyards are considered the top of the line, with Petite Champagne (a different type of chalk) a close second. After these, hierarchically, run Borderies (with more clay and sand), Fins Bois and Bons Bois (with varying amounts and types of chalk in their soil), then Bois Ordinaires (with more sand).

It used to be that only the top two or three *crus* were made into the top-priced cognacs, but now producers are experimenting successfully with finely aged cognacs blended from the other *crus* as well. Recently, Camus came out with a line of Île de Ré cognacs which have salty, peaty, whisky-like smoky aromas and flavours. It is particularly fitting that this line of cognacs may expose more people to cognac, because the superior salt harvested on the Île de Ré was a factor in early Dutch trading leading to the creation of the spirit of cognac itself.

Only white grape varieties may be used to make cognac. The main grapes in Cognac are Colombard, Folle Blanche and Ugni Blanc. In addition, cognac producers may use up to 10 per cent of Folignan, Jurançon Blanc, Meslier Saint-

François, Montils, Sélect, and Sémillon (each representing a maximum of 10 per cent). Cognac producers are already thinking about which grapes will do best in upcoming years if climate change progresses quickly.

Cognac must be produced with double distillation, and it must be aged in the fine oak barrels coopered in the region, with wood from nearby forests. Local Limousin oak, with its looser wood grain, is considered best for extracting the most desirable characteristics for brandy. The trees are carefully chosen, then the wood is cut, split and air dried for two to three years.

These costly barrels are constructed by hand. Each cognac producer may decide how the inside of the barrels will be toasted: some prefer it lighter and some darker, depending on the sensory features they want to impart to their spirits. (Though cognac ideally gains most of its colour from ageing in barrels, legally, some caramel colouring may be added, as well as sugar and a wood 'tea' called *boisé* for the final blending.) When cognac ages naturally, it moves from clear through yellow then tan into golden amber and tawny colours, and at ten years old it will be a rich, mahogany brown.

Through it all, producers must take into account evaporation, which ranges from 2 to 6 per cent per year. This evaporated amount is called the 'angels' share'. In French it's *la part des anges*, which is also the name of Cognac's annual charity auction, one which attracts high-end international bidders.

Cognacs are aged in both dry and humid cellars, sometimes being moved every year or so (typically just the contents, not the barrels) in a complicated scheme developed to maximize desirable aromas and flavours. In humid cellars the alcohol evaporates faster and the cognacs become rounder, softer, with more fruit and floral characteristics. Dry cellars have relatively faster water evaporation, and these conditions contribute more spicy and woody notes to the cognacs.

Distillers and cellar masters each have their own individual formulas for generating the particular taste profiles of their cognacs. These sensory elements range from floral through citrus, to baking spices and cigar-box-like aromas, and include innumerable other components like toffee, coffee, cedar, leather, dried fruits and vanilla.

Even after cognac's production methods were standardized, the rest wasn't quite smooth sailing. During the Second World War much of the area was occupied, but the far-sighted cognac producers were able to preserve much of their stocks for the future. Just how this was accomplished is not a story that is commonly told to visitors because it is said that some of the producers may have made deals with the devil; for those interested in the specifics there are more details in lengthier books on cognac's history.

The Cognac Bureau (the BNIC or Bureau National Interprofessionnel du Cognac) was created after the Second World War, in 1946, to monitor cognac's production and distribution both at home and throughout the world. Its plenary board consists of seventeen grape growers and seventeen cognac houses. According to BNIC rules, grapes for cognac must be made into a wine that is not aged; the wine must be distilled into an *eau de vie* as soon as possible. In Cognac, the distillations must be finished by 1 April of the year after the harvest. A cognac officially cannot be sold to the public until it has been barrel-aged for at least two years after the 1 April distillation limit.

One well-known cognac house, Hine, prides itself on continuing the practice of making a style of cognac that was very common until the mid-twentieth century. Called 'early landed' cognac, it is shipped to Britain in barrels and aged in a warehouse there before bottling. In England's year-round cool and damp climate, less evaporation occurs and the flavour profile is somewhat different from Cognac-aged spirits. (Hine has also been the official cognac supplier to Queen Elizabeth II for the past 50 years.)

It is important to note that all of the development of a cognac occurs during its barrel-ageing: once bottled, the cognac is ready to drink. It does not further improve in the bottle. Even in the barrel, at some point a cognac ceases to develop. This takes decades, up to as much as 80 years. At that point, if the cognac is not being bottled for sale, the cellar master will transfer it to a large, rounded glass container called a *dame-jeanne* (demijohn) and place it reverently in the gated and locked room that guards the producer's most precious cognacs. This area of the cellar is called, appropriately, the *paradis* (paradise).

Once opened, a bottle of cognac will keep for months or even up to a year in a cool, dark location. The best phase for drinking a cognac may be only a few years after bottling for a lesser-aged cognac, while it can stretch to decades for an older one. Though eventually even a bottled cognac's flavours and aromas will begin to fade.

Requirements for current cognac styles were finalized in 1983; and though future changes are being discussed, they are not guaranteed. Currently, these are the categories listed on the Cognac Bureau's website: VS (Very Special) or 3-star is *compte 2*, with at least two years of barrel age; VSOP (Very Superior Old Pale) or Reserve means *compte 4*, or at least four years of barrel ageing; Napoléon, XO (Extra Old) or Hors d'âge indicates *compte 6*, or at least six years of barrel-ageing. The category of the cognac reflects the age of the youngest brandy in the blend. Vintage-category cognacs, which have recently become popular, must contain spirits distilled from only one designated harvest. Most of the cognac sold (about 85 per cent) is VS or VSOP, with the rest being older cognacs.

There has been some confusion about the term Napoléon because it has been used indiscriminately in the past by producers in Cognac (and in the rest of the world) to imply that a brandy is very old. In Cognac, the term Napoléon now officially means the cognac has been aged at least six years, the same as an XO cognac. To further complicate things, in practice a cognac labelled with the word 'Extra' is often older than an XO. Many XO and Extra cognacs have a significant percentage of much older cognacs blended in; the exact components in the blend are different at each cognac house. Lately, cognac houses have also been releasing speciality cognacs with proprietary names and distinctive labels to capture the attention of different types of consumers.

Designations like VS, VSOP and XO have been co-opted by many countries, even those that no longer use the term 'cognac'. This labelling is convenient because it can—if employed honestly—tell buyers whether to expect a more- or less-aged product, within that particular system. However, each brandy-producing region has its own rules for ageing and labelling, so a VSOP from another region may not be the same age (or made with the same care) as a VSOP from Cognac.

At present, the Cognac Bureau lists 325 cognac producers, including small and large companies, individuals and cooperatives. But four major companies eclipse most of the others in that their names are practically household words worldwide: Courvoisier, Hennessy, Martel and Rémy Martin. They are responsible for about 85 per cent of Cognac's production. These companies own some vineyards, they distil some cognac from wine supplied by grape growers, and they also subscribe to the common practice of buying distilled spirit from growers and then ageing, blending, bottling and marketing the cognac.

Bottles these days range from cylindrical to flask-shaped, but there are many other curvy, seductive glass forms and contours. Producers commission amazing, sculptural shapes for their top cognacs; at this level the bottle itself adds significantly to the value of the cognac. Labels also vary widely in motif. They can appear ancient and traditional or be so completely postmodern that there are barely two words on the front of the bottle. All in all, design can add so tremendously to the prestige and price of a cognac that a few top-of-the-line bottles sell for thousands of dollars—or even tens of thousands. Single bottles of Hennessy, Hine and Courvoisier, for example, have sold for U.S. $10,000 (£6,000) in the past five years.

When exploring the mysteries of a fine spirit, it is tempting to start with the least expensive or youngest. This is not a good idea with cognac. A great place to begin is at the VSOP level, because here it is possible to experience the finesse of the aromas as well as the smoothness of the spirit—without spending a fortune. Even when planning to use cognac in a cocktail, mixologists say VSOP is still the best level to start with, as it melds surprisingly well with mixers and other flavourings.

VSOP cognac prices start around £30 ($45) and go up to some heights from there. Of course, VS cognac can be had for less. Most reputable brands of XO start in the neighbourhood of £60 ($100) for the top houses, while lesser-known (sometimes less reputable) cognac houses may price their XOs considerably lower.

When looking to buy a bottle of cognac, in general, price matters. But in addition to VS, VSOP and XO cognacs, there are also diverse cognacs being produced with proprietary names, and some are on the shelves for only a short time, increasing their appeal through exclusivity. Most of these special-release cognacs are fairly pricey, and what they offer—in addition to a uniquely designed label and/or bottle—is a different take on cognac in terms of aromas and flavours. One company might emphasize spicy notes while another may concentrate on blending more fruit flavours into the spirit, depending on the customer they want to attract.

In relation to drinking cognac, many people have seen the old caricature of a wealthy gentleman with his giant, afterdinner balloon glass sporting a small measure of tawny liquid at the bottom. In fact, this type of large, round-bowled glass is rapidly going out of style. It was useful 50 or 100 years ago, when houses were kept at a much cooler temperature than they are today and brandy would typically be stored in an even colder cellar. In order to liberate the delicate aromas and flavours of the cognac, people were encouraged to use the warmth of their hands to heat the liquid in the glass. As they heated, aromas were gradually released and contained for some time in the top of the balloon-shaped glass, where they could be experienced by gentle inhalation.

We no longer need to do this. In modern houses, cognac stored at room temperature starts out warm enough to drink. After the cognac is poured, the aromas are ready to be encountered by immediately inhaling at the rim of the glass. In addition, swirling the liquid around to release the aromas is not encouraged: a gentle rotary motion of the glass is all that is needed to release the delicate fragrances of a warm cognac.

In Cognac, it has become fashionable to use glasses that are similar in size to small white wine or sherry glasses. (This is almost a return to an early style of cognac glass, which was tiny, probably crafted more for imbibing than lengthy warming and appreciation.) Aged, sipping cognac should be served in small amounts: 20–40ml, or around 1 fl. oz.

When shopping for cognac, it is easy to see that the top four cognac houses, with 85 per cent of Cognac's production, dominate sales around the world. Most of them have large corporations behind them, providing resources for global expansion, product development, distribution and marketing. In part or in total, LVMH Moët Hennessy Louis Vuitton owns Hennessey, Pernod Ricard owns Martell, Beam Global (now itself owned by Suntory) owns Courvoisier and Rémy Martin is owned by the Rémy Cointreau company. In March 2012 these four companies were reported to have generated around £3 billion ($5 billion) in worldwide sales during the previous year.

There are many other fine cognacs to try, depending on availability, the occasion—and one's bank balance, of course. These additional brands tend to be the most widely available: ABK6, Bache-Gabrielsen, Baron Otard, Bisquit, Camus, Conjure, De Luze, Delamain, Ferrand, François Voyer, Frapin, Gautier, Hardy, Hine, Jean Fillioux, Jenssen, Landy, Louis Royer, Meukow, Normandin-Mercier, Peyrat and Prunier.

Chapter 10
Cognac Cocktails and 21st-century Trends

Luxury is the essence of cognac's image in the twenty-first century. Whether cognac is sipped straight, blended into a high-end cocktail or combined with a mixer, the people imbibing it are all pursuing a deluxe experience. In Britain, the U.S. and countries that orient themselves culturally with the West, there is still a reliable segment of the population interested in high-end brandies that are sipped neat, after dinner. Cognac's luxury lifestyle association has remained important at this level, with top-tier cognacs dominating. However, the current brandy renaissance has several trending sources: East Asian culture, American popular music and lush mixology.

Two concurrent cultures began consuming a great deal more cognac and brandy at the end of the twentieth century. One was in Asia, specifically high-end Hong Kong consumers. The other was urban (inner city) U.S. consumers. With cognac growth basically flat in most other parts of the world, cognac producers began paying much more attention to these two areas, and they have definitely reaped the rewards.

Increased urban U.S. brandy and cognac consumption occurred in locations with predominantly Hispanic and black populations. Within Latino society, this could be the result of further extension of the Spanish brandy culture already prevalent in Mexico and other Latin countries. Within urban black society, a number of explanations have been suggested, from rebellion and differentiation of current young people to the European experience of black servicemen in the Second World War—though the latter cause would date any increased interest in brandy pretty far back in the twentieth century. In fact, there had been considerable consumption of brandy in the inner city for several decades before the urban rap movement began to vocalize it. Much of the rest of the world only became aware of it in 2001 with Busta Rhymes's song featuring P. Diddy, 'Pass the Courvoisier', and with its subsequent (equally successful) remix and music video.

In the dozen years after that, over 150 rap songs were released containing references to cognac brands, especially Courvoisier and Hennessy (also spelled Hennessey). At the same time, to concentrate on this market, cognac producers formed partnerships with several major artists. Ludacris put out his Conjure cognac with Kim Hartmann, owner of Birkedal Hartmann cognac—though in the advertising campaign Ludacris is identified only by his real name, Chris Bridges. Dr Dre released his Aftermath cognac in conjunction with ABK6 cognac. The rapper T.I. and Martell cognac announced their affiliation—but Martell dissolved it a few months later when T.I. was sent to prison.

Cognacs have also been produced specifically for the high-end urban trade by some of the most important cognac companies. The musician Jay-Z introduced d'Ussé to the U.S.; this cognac is a product of the Bacardi Company, which also owns Baron Otard cognac. Rémy Martin recently had a limited release of a VSOP called Urban Lights, with a label design that glowed red in ultraviolet light. And there are more each year. It is interesting to note that on their websites cognac companies do not always list these products, preferring that consumers visit the websites with the proprietary names of the particular cognacs. At this point in time, cognac producers seem to be following several different paths simultaneously in their quest to sustain the global luxury appeal of this classic spirit.

Turning to Asia, there is only one word for cognac consumption: explosive. Though many people have heard that top Bordeaux wines are now requisite accessories for the high-end Chinese consumer, not everyone knows that cognac is now the Bordeaux equivalent in spirits. Whereas Japanese customers prefer Scotch whisky as their favourite Western spirit, the Chinese and other Asians are fervent consumers of brandies, and in China that means XO, Extra and Vintage: only top of the line will do. The older and rarer, the better.

In 2012 the Chinese market for cognac outstripped the erstwhile number-one market, the United States, in value. Due to this gargantuan growth rate, much of the focus of cognac's producers has been on Asia for the past few years. And because the Chinese tend to buy only the top of the line, this situation is expected to continue for the foreseeable future. A sizeable number of cognac houses is producing special bottlings, blends and labels for the Chinese market—whether or not this is something they publicize in the rest of the world.

Recently, articles in the press have questioned whether cognac can continue to satisfy Chinese demand. Savvy importers and consumers in China have also discovered armagnac, to the Armagnac region's great delight. Armagnac's noble brandies were second-in-line for centuries due to its landlocked location, but that obstacle has been overcome by modern transportation. Armagnac now basks in well-deserved prestige in China.

But the Chinese are reaching even further now, to additional markets that make authentic (grape-wine-based) brandies, the closest being Australia. For new markets, sourcing brandy from areas other than Cognac is fairly simple because most brandy producers around the world label their brandies in the same way: 3-star or VS is generally the first level, VSOP the second level, and XO the third. In addition, terms like Napoléon, Extra Old and Vintage are used there, as well as proprietary names for special releases. Though the brandies in each category are not uniformly aged across the world, this labelling makes it easier for new distributors and consumers to get their start in the brandy world.

Other Asian countries are going along for the ride, too. In some, their brandy customs began with early colonial influence. In others it has come straight from urban life as seen in music videos—which is the case among young people in the Philippines today. However, other generations in this country have also maintained a tradition of consuming Spanish brandy, a favourite being Pedro Domecq's brandy de Jerez, Fundador.

There is a certain amount of cognac imported into Asian countries other than China. Vietnam, for instance, was the fifth-largest brandy consumer in the late 1990s, and its brandy consumption remains strong today. Brandy drinkers have mainly been men in the 20–50 age range, with their preferred spirit being cognac; other brandies are chosen by less than 10 per cent of these males, with about 1 per cent favouring armagnac. Top cognac houses are preferred, since much of this spirit is consumed publicly by young men in venues like nightclubs.

Brandies are also being produced in Russia and in other Asian countries. However, many of these are not actually grape wine distillations, as there are not many grapes grown in some of these regions. Brandy is big in Malaysia—in fact the biggest brandy brand in the world is Malaysia's Emperador. But the key here is what is *not* said: what the brandy is made from. Only sometimes is it made from grapes, and in those cases only sometimes are the grapes wine grapes (as opposed to table grapes).

However, there is a trend towards authenticity in certain markets. Russia's Fanagoriysky now emphasizes the oak ageing of its brandies, as do many other brandy producers on this continent. Fanagoriysky recently opened its own modern cooperage. Russia's KiN Group owns the Domaine des Broix in Cognac, from which they import distilled spirits and aged cognac. KiN also makes domestically produced, blended brandies. India's Morpheus uses some grapes grown in India and some imported from France. China is making attempts at traditional-style brandy distilling, according to a consultant who works there.

In various places, 'brandy' is simply a label on a bottle of a coloured distillation of whatever agricultural product it makes sense to use there—even pineapple. The advantage is the lower cost for the producer, as well as the customer; these brandies sell for less than half the prices of the imported brandies and cognacs.

In terms of brandy products, there is one more curious trend to mention, which seems to be happening everywhere from Armagnac to California: 'white' brandy. In the spirits industry 'white' actually refers to clear spirits. For brandy, this would seem to be a contradiction in terms, as brandies are known for their wood ageing which imparts wonderful aromas, flavours and colours to the spirit. White brandies can appear clear because they are unaged, or because they have had the wood tint filtered out.

White brandies are also native to Armagnac—in fact, their traditional 'blanche' received its official AOC (Appellation d'Origine Contrôlée) title in 2005. Blanche must be made from designated vineyard parcels, from the grapes Folle Blanche, Ugni Blanc, Baco and Colombard. Early distillation is required, followed by a three-month settling or 'maturation' period. Then a producer can begin adding water to the high-alcohol spirit to bring it down to its bottling level of around 40 per cent. In practice, this phase usually takes place over a period longer than three months.

White brandies capitalize on the fashion for clear spirit bases for cocktails, a trend with notable growth the last few decades which does not seem to be abating. The white brandy fashion is visible even in cognac. In 2010 Rémy Martin produced a clear 'V' unaged spirit from cognac. Hennessy, which also has a Black cognac, did a 'Pure White', which was tested in several countries. Technically these cannot be called 'cognac' because cognac requires barrel-ageing for a specified time. But more white brandies will probably be available soon, in part because this could be a way to sell brandy from the Cognac region without waiting the many years required for ageing traditional styles.

In other areas, South Africa's Collison's makes a White Gold brandy. In the U.S., Christian Brothers has jumped into the ring with its 'Frost' brandy. This is aged in oak, but then processed (and flavoured, in this case) so it appears clear in the bottle. For all these clear spirits, the recommendations are to drink them chilled—perhaps as an alternative to vodka in cocktails for the fashionable mixology trend. And it is in the mixology arena that cognac is really getting hot.

Chapter 11
Small-batch Brandies and Cognacs

Even before the cocktail culture began revitalizing cognac early in the twenty-first century, there were stirrings of brandy's revival around the globe, particularly in the U.S. where, in the 1980s, the artisan spirit movement encouraged new brandy producers to get into the game. But of the two most famous new producers, one has become a staple of artisan brandy in the U.S., and one has disappeared.

Cognac's illustrious Rémy Martin cognac company formed a partnership with Jack Davies of California's Napa sparkling winery Schramsberg Vineyards in 1982. They established a distillery in the Carneros wine region that straddles Napa and Sonoma. It was called RMS, and great things were expected. But rather than being lauded for their authentically cognac-styled brandy, RMS received little attention from American customers—even when they later changed their name to Carneros Alambic.

Also during the early 1980s, the distillery Germain-Robin was founded in California's remote, wooded, northern county of Mendocino, by the American Ansley Coale and Cognac native Hubert Germain-Robin. Again, there was little excitement in the U.S., though the brandies received consistently great reviews. Both companies persevered for more than fifteen years. But by the late 1990s, RMS/Carneros Alambic had folded, while Germain-Robin had finally turned a profit and has continued producing artisanal spirits to this day.

Coale and Germain-Robin instituted a new approach to selecting grapes for their brandies. At first they sourced the same grape varieties that are used in Cognac. But Coale and Germain-Robin discovered that the brandy made in Mendocino, California, did not have the taste they were after, because the grapes were not grown on Cognac's chalky soil. Realizing they were in the midst of excellent wine grape country, they decided to take a chance and experiment with making their brandies out of the best local grapes. After exhaustive trials they eventually discovered how to make fine-quality brandy with these grapes—though Coale and Germain-Robin did have to persuade the farmers they contracted with to pick a little earlier to maintain the acidity balance their brandy needed. Now Germain-Robin uses one cognac grape, Colombard, but relies mainly on Pinot Noir. Depending on the harvest, they also include locally grown grapes such as Sémillon, Sauvignon Blanc, Zinfandel, Chenin Blanc and Muscat in their brandies.

Once they had chosen the best ingredients, Coale found that he had to educate the public about the contents of his brandies. He observed that Americans looked at brandy as a kind of static commodity: they bought by brand name and moved up by price, not really aware of what went into the brandies. Slowly, Coale has been taking the American public along a learning curve.

After his success in Mendocino, the distiller Hubert Germain-Robin has gone on to consult with brandy producers in Asia and other parts of the world. Because of the popularity of cognac and brandy today, many companies are eager to become brandy distillers. But not all of them have good grapes to distil with—and some don't even have grapes—so Hubert often finds himself in the midst of a fascinating exploration of this brave new world.

Back in the U.S., a few other producers had decided to jump into the game. Daniel Farber of Osocalis in the Santa Cruz area of California calls himself 'generation 1.5' of California brandy. Before establishing Osocalis, he travelled to France and Spain to learn about brandy production and ageing. When he started distilling later in the 1980s, he was not aware of the new RMS and Germai-Robin distilleries; he simply wanted to produce a world-class brandy. Now, he credits Hubert Germain-Robin with being a visionary and pioneer in creating the 'California alembic' style of brandy.

Farber uses cognac-type distillation and California grapes such as Pinot Noir, Sémillon and Colombard; every year the blend is a little different. Stylistically, Farber believes Osocalis brandies are a cross between armagnac and cognac. He tried ageing his brandies in American oak, but realized French Limousin oak was far better. Though he usually releases each vintage when it has aged (and doesn't generally blend them), he is enthralled by the notion that a brandy and a human have roughly the same age span: up to 80 years.

Jepson is another California brandy that began in Mendocino. Bob Jepson bought a property near the Russian River in 1985 to establish a distillery and winery there. The current owners took it over in 2009 and named it Jaxon Keys Winery and Distillery. Under the Jepson name they still produce brandy that is distilled on the property and made from the Colombard grown in their vineyards.

Back in the Carneros region, when the Etude winery took over the former Carneros Alambic property (originally RMS) in 2002, they also acquired brandy that was ageing there in casks. The company now sells an expensive Etude XO brandy—aged twenty years and blended by their winemaker—which was made from locally grown grapes: Pinot Noir, Colombard, Chenin Blanc, Palomino, Chardonnay, Ugni Blanc, Muscat and Folle Blanche.

Another distillery that started as a small artisanal venture is Charbay, which boasts a thirteenth-generation distiller whose father immigrated from the Balkans in 1962. This family enterprise began in 1983 on Spring Mountain, part of the range that divides the Napa and Sonoma wine regions. After great success with their vodkas, the Karakasevic family branched out and a few years ago they released their Brandy No. 83, made from Folle Blanche grapes—a spirit which they had been aging for 27 years.

Undoubtedly, even as this chapter is being read, small distillers in various parts of the country are starting up new brandy ventures. One of the most recent is Finger Lakes Distilling in rural upstate New York, which is owned by Brian McKenzie. He also began by producing other grape-based spirits, as his distillery is located in the middle of the Finger Lakes wine region. McKenzie is especially interested in brandy because so few artisan producers are making it. He distilled the brandy soon after the company started up a few years ago. The first batch was released in 2012 when it had aged to his satisfaction; it was made with local grapes including Gewürztraminer and other native and hybrid varieties.

Another notable young distiller, Emmanuel Painturaud, is located in Cognac itself. He is part of a small, intriguing initiative spearheaded by a small number of families who have been growing grapes and making wine to supply the large cognac houses. But as of a few years ago, a tiny percentage of grape growers in Cognac has begun to take on the extra financial risk of ageing, labelling and marketing their own cognacs. They are hoping that with an increased interest in cognac and in all things artisanal, the public will be receptive to a family-made cognac. This movement mirrors the initiative of the 'grower-producers' of the Champagne region, who began receiving outside attention—and great reviews—just about the turn of the twenty-first century, and are now a great success.

Emmanuel works with his father and brother on a small farm where his grandfather first had a modest still in 1934. They grow grapes, make wine, and distil and sell some to

Rémy Martin, which requires them to distil wine on the lees for 'a rounder, deeper richer flavour'. And they are proud of the product they supply to this large company. But now, Emmanuel also takes some of his own wine and distils it to make his own artisanal cognacs.

The Painturaud family uses their older barrels for ageing and storing the cognacs they blend for themselves. They buy four or five of the very expensive new barrels a year. These can be considered 'new' for the three years, which means that they can be used for the critical first six months of cognac ageing. The Painturauds blend three or four cognacs together to make their final blend of each style of cognac. In the past, the family aged and bottled a small amount for themselves. Now they are bottling to sell, so they make several different styles. Before Emmanuel came back to the family business, the Painturauds did not make an XO, but today they do because it is in demand.

There are about 1,000 family-owned operations in Cognac, but this number is decreasing as the adult children do not want to remain out in the countryside, working the land. While there are other young people who might want to start a small business of growing and distilling in Cognac, unfortunately they usually cannot afford it. Emmanuel blames this on the increase in land prices; he maintains that in the past five years vineyard prices have become nearly prohibitive in all areas of Cognac due to an enormous increase in demand fuelled by Chinese high-end customers and American rap musicians.

Another way family-aged cognacs come on the market is through a company like PM Spirits, owned by Nicolas Palazzi, who now lives in New York. Raised by his grandparents in Bordeaux, Palazzi established a custom, small-batch cognac business in 2008. Beginning with the cognac that a friend of his grandfather wanted to sell, Palazzi was soon referred to others in similar situations. Perhaps that family was getting out of the cognac business; perhaps they needed money; perhaps they simply wanted to take advantage of the current popularity of cognac around the world. Whatever the reason, Palazzi has been able to source aged cognacs, which he collects—and ages further, if necessary—then bottles and sells to elite customers in New York and other cities.

Palazzi also takes this one step further: he will custom-blend and custom-bottle cognac for individuals who desire their own distinctive cognac. After numerous careful, detailed tastings with a customer (at home or in Cognac) Palazzi works to design labels and distinctive hand-blown bottles for these unique blends of cognac, which are supplied to collectors and for special events.

There's a popular song called 'Everything Old is New Again'—and there is no truer place to apply this phrase than to the world of brandy today.

Recipes

Here are some of the most famous cocktail recipes that use cognac or brandy as their base spirit—and some new ones. Classics may have different proportions than the present-day cocktail imbiber is used to. Some classic cocktails are modernized here with slight variations, such as using brandy de Jerez instead of cognac. Cocktail connoisseurs will relish the opportunity to widen their realm of experience by tasting these cocktails.

Brandy Alexander

This is the first drink anyone mentions when asked for the name of a cocktail made with brandy. But nowadays few of us know what it is—except for mixologists, of course. From the 1930s to the 1970s, this was a very popular drink in the U.S., with some geographical differences: in the south, brandy was a man's drink, while in the north the frothiness of this cocktail made it more of a ladies' drink. This recipe comes from E. & J. Gallo, who would have produced the brandy used for many of these cocktails made in America during the mid-twentieth century.

30 ml (1 fl. oz) brandy
30 ml (1 fl. oz) cream
30 ml (1 fl. oz) dark crème de cacao
ground nutmeg

Shake liquid ingredients well with ice. Strain into a cocktail glass and dust with ground nutmeg.

Brandy Collins

One of the newer bodegas in Jerez is, ironically, called Bodegas Tradición. They source older brandies from around the region, then age and blend them before bottling. As a young company, they are open to new styles of cocktails with old brandies. Here is a recipe for a 'Collins' made with brandy de Jerez.

50 ml brandy de Jerez
30 ml fresh lemon juice
20 ml simple syrup
iced soda water
slice lemon

Mix all ingredients but the soda water with ice. Blend in a shaker, strain into a Collins glass filled with ice. Top with soda water, garnish with a lemon slice and serve with a straw.

Brandy Crusta

This is a recipe made for me by Alexandre Lambert,

bartender at Bar Louise in the Hôtel François Premier, which opened in 2012 in the centre of the city of Cognac. The cocktail is based on the original created by Joseph Santini at his bar The Jewel of the South in New Orleans in the mid-nineteenth century. Lambert says Hennessy's Fine de Cognac is a cross between a VS and VSOP; he also recommends using VSOP cognac for this drink.

1 dash Peychaud's Bitters

1 dash Angostura bitters

1 tsp original recipe triple sec

1 tsp Luxardo Maraschino liqueur

40 ml Hennessy 'Fine de Cognac'

2 tsp fresh lemon juice

1 tsp simple syrup

granulated sugar

lemon peel

Stir ice cubes in a glass jug. Add liquid ingredients and continue stirring until blended. Strain into a coupe glass rimmed with granulated sugar. Drop in a thin spiral of lemon peel.

Cognac Punch

The British created punch using brandy, the word 'punch' coming from the Hindi word for five, the number of ingredients used: sugar, brandy, lemon or lime juice, water and flavourings. Wine was sometimes used in place of, or in addition to, water. This recipe is from the Cognac Bureau (BNIC) in France.

peel of 4 lemons

250 g (9 oz) icing (superfine) sugar

250 ml (1 cup) fresh lemon juice, strained

750 ml (1 bottle, or 3 cups) VSOP (or VS) cognac

250 ml (1 cup) rum

1.5 l (6 cups) cold water

1 whole nutmeg

Muddle the lemon peels with the sugar and let sit for at least 1 hour. Muddle again and add the lemon juice, stirring until sugar has dissolved. Strain out the lemon peels. Add the cognac and the rum and stir. Refrigerate.

To serve, pour into a punchbowl filled halfway with ice cubes, add the cold water and stir. Grate 1/2 nutmeg over the top.

Cognac Long Drink

Visitors to Cognac in the summer are first surprised to see precious cognac diluted with soda or tonic. Once they have one in hand, they get it. The Cognac Long Drink is always in fashion during Cognac's hottest days and nights. For a simple 'long-drink', VS is generally used–but VSOP is even better. This recipe is from the BNIC .

30 ml (1 fl. oz) cognac

90 ml (3 fl. oz) tonic water

Pour into in a Collins glass containing ice cubes. Stir lightly.

French Connection

An after-dinner drink, with the amaretto not-so-subtly flavouring the cognac. It's also possible to adjust the proportions to taste, so that there is half as much amaretto as cognac.

30 ml (1 fl. oz) cognac

30 ml (1 fl. oz) amaretto liqueur

Pour over ice cubes in a highball glass and stir until blended.

Horse's Neck

A classic drink to sip on a warm evening while wandering through the Cognac Blues Passion, an incredible outdoor music festival held every July in a park in the centre of the town of Cognac. Here is the way it should be served, according to the BNIC.

30 ml (1 fl. oz) cognac, VS or VSOP

dash Angostura bitters

ginger ale

orange peel (optional)

Pour cognac over a couple of ice cubes in a Collins glass. Add a dash of bitters. Top up with ginger ale. Optional: add a thin twist of orange peel to decorate.

Hot Toddy

Essentially, this is lemony hot water sweetened with honey, with a 'medicinal' dose of brandy. For a little more flavour, make it with tea instead of hot water. Either way, people have been drinking this for centuries to ward off chills and colds.

1 tbsp honey

2 tsp lemon juice

125 ml (½ cup) very hot water

30 ml (1 fl. oz) brandy

Combine honey, lemon juice and hot water in a mug. Stir in brandy just before serving.

Medicinal Brandy

This can be given in the case of colds or scratchy throats. In the past it was also considered suitable for invalids or those recovering from an illness because the milk provides nourishment. It is traditional in Italy, as well as many other countries.

180 ml (6 fl. oz) milk

30 ml (1 fl. oz) brandy

1 tsp sugar (optional)

Heat milk to just below boiling—or use an espresso machine or frothing device to foam the milk. Pour into a glass or mug. Stir in brandy. Sweeten with sugar if desired. Serve either morning or night.

Modern Mojito

A recipe from Spain, from Pedro Domecq, makers of brandy de Jerez. This is an informal use for a traditional brandy, the favourite of one of the employees there. It was provided with no specific amounts; this recipe was derived by testing various combinations of the ingredients. It is meant to be a fresh and healthy drink, so feel free to add more mint leaves and lemon juice.

fresh mint leaves

2 tsp sugar

1 tbsp fresh lemon juice

50 ml brandy de Jerez

crushed ice

Muddle the mint leaves and sugar in a shaker. Add lemon and brandy and shake well, until sugar has dissolved. Pour into a Manhattan glass half-filled with crushed ice.

Pisco Punch

The original recipe—made by dedicated mixologists today—contains a syrup made from gum arabic from the acacia tree. There's a lengthy process involved in combining the gum arabic with a sugar solution, but mixologists swear it's worth it. The recipe below, provided by Pisco Portón from Peru, is a simplified version, easier for home entertaining.

1 fresh pineapple

235 ml (1 cup) simple syrup

470 ml (2 cups) bottled water

750 ml (1 bottle, or 3 cups) Pisco Portón

300 ml (1¼ cups) fresh lemon juice

Cut a fresh pineapple in pieces about 1.5 by 4 cm (0.5 by 1.5 in.) and soak overnight in simple syrup. In the morning, mix the rest of the ingredients in a big bowl. Lemon juice or simple syrup may be added to taste. Use around 100 ml (3–4 fl. oz) of punch per glass, adding a cube of the soaked pineapple to each.

Pisco Sour

When there's a party in Peru, people expect Pisco Sours. According to the sisters who created Macchu Pisco, here's how they do it:

2 parts Peruvian pisco (preferably Quebranta-grape pisco)

1 part fresh lime juice

1 part sugar

1 shot of egg white

Angostura bitters

Ice

Put the first four ingredients in a blender with two cups of ice. Blend well. Pour into a chilled glass. Top each cocktail with a few drops of Angostura bitters. Number of servings? Your call . . .

Sidecar

From the BNIC, this recipe begins with a great history of the drink by the mixologist Dale DeGroff:

The Sidecar is our legacyfrom the Crusta, although the twentieth century chapter to the story is poorly documented. Harry's New York Bar claims creditfor the Sidecar. But Colin Field, the head bartender at the Ritz Hotel's Hemingway Bar in Paris, is convinced hispredecessor Frank Meier, Hemingway's legendary barman in the early days of the Ritz, created the drink sometime in 1923 although there is no documentation to prove his claim. The single bit of evidence we have inprint is the 1922 book by Robert Vermeire of the Embassy bar in London called 'Cocktails: How to Mix Them'. In it, the drink is credited to a barman named MacGarry at the

Bucks Club in London.

45 ml (1½ fl. oz) VSOP cognac

30 ml (1 fl. oz) triple sec

20 ml (¾ fl. oz) fresh lemon juice

orange zest (optional)

Combine all ingredients in a mixing glass and strain into a small cocktail glass with a lightly sugared rim. Garnish with a small flamed orange zest.

Stinger

This cocktail was a classic for many years, though it may not be the first on everyone's list today. The name probably refers to the 'zing' you get with this drink. According to personal preference, the amount of crème de menthe may be decreased; some people like the proportions of two parts brandy to one part crème de menthe.

30 ml (1 fl. oz) brandy

30 ml (1 fl. oz) crème de menthe

crushed ice

fresh mint leaf (optional)

Put in a shaker with crushed ice and shake until blended. Strain into a cocktail glass. Garnish with fresh mint leaf if desired.

Summit Cocktail

A few years ago, when the BNIC began encouraging bartenders to experiment with cognac in cocktails, they held a Cocktail Summit and invited numerous celebrated mixologists. This cocktail was created for the event by Andy Seymour and is now a new classic, served at many Cognac Bureau occasions around the world.

zest of 1 lime

4 thin slices fresh ginger

45 ml (1½ fl. oz) VSOP cognac

60 ml (2 fl. oz) lemon and lime soda

1 long piece cucumber

Put lime and ginger into a glass, pour in half the cognac, press lightly 2–3 times. Half fill with ice, stir for 5 seconds. Pour in remaining cognac. Add lemon and lime soda and cucumber, stir well and serve immediately.

Suprême de l'Armagnac

This cocktail was recently developed by Philippe Olivier, head barman at the Hôtel de Crillon in Paris, working with the Armagnac Bureau (BNIA). It melds citrus and brandy, which is a classic flavour combination that works for many palates around the world.

40 ml (1¼ fl. oz) armagnac

30 ml (1 fl. oz) grapefruit juice

1 tsp orange juice pressed from peeled orange segments

maraschino cherry (optional)

Shake all ingredients together well over ice. Strain into cocktail glass. Optional: garnish rim with a maraschino cherry.

Select Bibliography

Calabrese, Salvatore, *Cognac: A Liquid History* (London, 2005)

'Clem Hill Tell Test History', *Daily News*, Perth (11 March 1933), from http://nla.gov.au

Cullen, L. M, *The Brandy Trade under the Ancien Régime: Regional Specialisation in the Charente* (Cambridge, 1998)

Dicum, Gregory, *The Pisco Book* (San Francisco, 2011)

Faith, Nicholas, *Cognac* (Boston, 1987)

Fromm, Alfred, *Marketing California Wine and Brandy: Oral History Transcript*, ed. Ruth Teiser, Regional Oral History Office, The Bancroft Library, University of California at Berkeley (1984)

Jarrard, Kyle, *Cognac: The Seductive Saga of the World's Most Coveted Spirit* (Hoboken, NJ, 2005)

Kops, Henriette de Bruyn, *A Spirited Exchange: The Wine and Brandy Trade between France and the Dutch Republic in its Atlantic Framework, 1600–1650* (Boston, 2007)

Miller, Anistatia, and Jared Brown, *A SpirituousJourney: A History of Drink, Book One–From the Birth of Spirits to the Birth of the Cocktail* (Cheltenham, 2009)

——, *A SpirituousJourney: A History of Drink, Book Two—From the Publicans to Master Mixologists* (Cheltenham, 2010)

Neal, Charles, *Armagnac: The Definitive Guide to France's Premier Brandy* (San Francisco, 1998)

Wilson, C. Anne, *Water of Life: A History of Wine-Distilling and Spirits 500 BC–AD 2000* (Totnes, Devon, 2006)

Websites and Associations

Brandy Producers

Asbach Brandy

www.asbach.de

Beam

www.beamglobal.com

Birkedal Hartmann

www.birkedal-hartmann.com

The Blanche

www.theblanche.com

The Christian Brothers

www.christianbrothersbrandy.com

E&J Brandy

www.ejbrandy.com

Etude

www.etudewines.com

Finger Lakes Distilling

www.fingerlakesdistilling.com

Hine

www.hinecognac.com

Jaxon Keys Winery and Distillery

www.jaxonkeys.com

Korbel

www.korbelbrandy.com

PM Spirits

www.pmspirits.com

Proshyan

www.proshyan.am

Romate

wwww.romate.com

Brandy Information

All About Brandy, Cognac and Armagnac

www.tastings.com/spirits/brandy

The American Distilling Institute
www.distilling.com

Armagnac: The True Spirit of France
www.armagnac.fr

Beverage Media Group
www.bevnetwork.com

Brandy de Jerez
www.brandydejerez.es

Bureau National Interprofessionel de Cognac
www.cognac.fr/cognac

Cognac Expert
www.cognac-expert.com

Drinks International
www.drinksint.com

Experience Cognac
www.experience-cognac.fr

International Bartenders Association
www.iba-world.com

Paul Masson Winery Operations and Management, 1944—1988
http://archive.org

Pisco Chile
www.piscochile.com

Derek, Ramsden, *South African Brandy in an International Context*,
Dissertation, 2012
www.capewineacademy.co.za

Rap Genius
www.rapgenius.com

Gary, Regan, 'Behind the Drink: The Brandy Alexander'
http://liquor.com

Shanken News Daily
www.shankennewsdaily.com

South Africa Brandy Foundation
www.sabrandy.co.za

Acknowledgements

In Armenia: Ararat/Pernod-Ricard, Noy, Proshyan Brandy, Vedi-Alco; in Armagnac: Amanda Garnham, the BNIA and all its members, Ithier Bouchard at Tariquet, Jean Castarède, Arnaud and Denis Lesgourgues at Château de Laubade; in Australia Matt Redin at Angove, Rob Hirst at Fine Wine Partners, John Geber and Marty Powell at Château Tanunda, in Cognac: Jean-Louis Carbonnier, Nicki Sizemore and the BNIC and all its members; in Georgia: Tina Kezeli and Georgie Apkhazava of the Georgian Wine Association Ekaterine Egutia, Zviad Kvlividze, David Abzianidze and Sarajishvili, Tfilisi Marani, Kakheti Traditional Winemaking; in Jerez: Beam Domecq, Carmen Aumesquet and Cesar Saldaña at Consejo Regulador del Brandy de Jerez, Bodegas Fernando de Castilla, González Byass, Sánchez Romate, Lorenzo Garcia-Iglesias at Bodegas Tradición; for pisco, Elizabeth and Melanie Asher at Macchu Pisco, Johnny Schuler at Pisco Portón, Guillermo L. Toro-Lira; in South Africa: Louise and Tessa de Kock, Elsa Vogts KWV; in the U.S., Bill Owens at the American Distilling Institute, Scott DiSalvo and Russell Ricketts at E&J Gallo, Brian McKenzie at Finger Lakes Distilling, Ansley Coale, Hubert Germain-Robin, Dan Farber at Osocalis.

Additionally: David Baker, Tim Clarke, Concord Writers Group, Jill and Dale DeGroff, Pierluigi Donini, Branko Gerovac, Kyle Jarrard, Lauren Kinelski at RemyUSA, Sandra MacDonald, Charles de Bournet Marnier Lapostolle Maria Mata at Mascaro, Elizabeth Minchilli, Mark Pallot at Beam Global, Norm Roby, Ken Simonson, Hamish Smith, Jan Solomon, Calvin Stovall, Ann Tuennerman at Tales of the Cocktail, Elisa Vignuda and Kirsten Amann at Fratelli Branca, Rosie Vidal, David Wondrich.

Photo Acknowledgements

The author and the publishers wish to express their thanks to the below sources of illustrative material and/or permission to reproduce it.

Courtesy of Angove: pp. 078, 080; Courtesy of Armagnac Delord: pp. 037, 039 top, 041 bottom, 043; Bigstock: p. 2(contents page)(Marco Mayer); © BNIC: pp. 5, 100 (Roger Cantagrel), 011, 019, 107 (Gérard Martron), 018, 030 (Bernard Verrax), 096 (Jean-Yves Boyer), 103 (Stéphane Carbeau); Bureau National Interprofessional de l'Armagnac (BNIA): p. 015; Courtesy of Chateau de L'Aubade: pp. 032, 033, 034, 035, 037, 039, 041, 042, 043 (Michael Carossio); Courtesy of Chateau Tanunda: p. 076; Courtesy of Chateau du Tariquet: pp. 032, 035; Courtesy of Cognac Hardy: p. 7; Courtesy of E&J Gallo: p. 089; Becky Sue Epstein: pp. 053, 071; Branko Gerovac: pp. 048, 049, 050, 051, 052, 053, 055, 056, 057 top, 060; Courtesy of Germain-Robin: pp. 115, 117; Courtesy of Gonzalez-Byass: p. 016; Courtesy of Janneau: p. 102; Courtesy of Korbel Brandy: pp. 089, 090, 091, 092; Courtesy of Nicolas Palazzi: pp. 120, 121; Courtesy of Sànchez

Romate: pp. 041 top, 063, 065, 066; Courtesy of Sarajishvili: pp. 057 bottom, 059; U.S. National Library of Medicine, Bethesda, Maryland: p. 8; image supplied by Van Ryn's: p. 081.

图书在版编目（CIP）数据

白兰地 ／（美）贝基·苏·爱泼斯坦著；陈媛熙译 .
-- 北京：北京联合出版公司，2023.6
（食物小传）
ISBN 978-7-5596-6827-1

Ⅰ.①白… Ⅱ.①贝… ②陈… Ⅲ.①白兰地酒 -
普及读物 Ⅳ.① TS262.3-49

中国国家版本馆 CIP 数据核字（2023）第 058880 号

白兰地

作　　者：〔美国〕贝基·苏·爱泼斯坦
译　　者：陈媛熙
出 品 人：赵红仕
责任编辑：孙志文
产品经理：夏家惠
封面设计：鹏飞艺术
封面插画：〔印度尼西亚〕亚尼·哈姆迪

北京联合出版公司出版
（北京市西城区德外大街 83 号楼 9 层　　100088）
北京天恒嘉业印刷有限公司印刷　　　新华书店经销
字数 118 千字　889 毫米 ×1194 毫米　1/32　8.75 印张
2023 年 6 月第 1 版　　2023 年 6 月第 1 次印刷
ISBN 978-7-5596-6827-1
定价：59.80 元

版权所有 侵权必究
北京市版权局著作权合同登记　图字：01-2022-5539 号